FLUID/STRUCTURE INTERACTION DURING SEISMIC EXCITATION

A report by the
ASCE Committee on Seismic Analysis of the Committee
on Nuclear Structures and Materials of the Structural Division

Published by the
American Society of Civil Engineers
345 East 47th Street
New York, New York 10017-2398

The material presented in this publication has been prepared in accordance with generally recognized engineering principles and practices, and is for general information only. This information should not be used without first securing competent advice with respect to its suitability for any general or specific application.

The contents of this publication are not intended to be and should not be construed to be a standard of the American Society of Civil Engineers (ASCE) and are not intended for use as a reference in purchase specifications, contracts, regulations, statutes, or any other legal document.

No reference made in this publication to any specific method, product, process, or service constitutes or implies an endorsement, recommendation, or warranty thereof by ASCE.

ASCE makes no representation or warranty of any kind, whether express or implied, concerning the accuracy, completeness, suitability or utility of any information, apparatus, product, or process discussed in this publication, and assumes no liability therefor.

Anyone utilizing this information assumes all liability arising from such use, including but not limited to infringement of any patent or patents.

Copyright © 1984 by the American Society of Civil Engineers,
All Rights Reserved.
Library of Congress No. 84-70058
ISBN 0-87262-392-0
Manufactured in the United States of America.

Membership For

Fluid-Structure Interaction Working Group

Seismic Analysis Committee

D. D. Kana, Chairman
Southwest Research Institute
San Antonio, Texas

D. V. Reddy
University of Florida
Gainesville, Florida

C. V. Subramanian
General Electric Company
San Jose, California

I. K. Ghosh
PAL Consultants, Inc.
Santa Clara, California

Q. A. Hossain
Quadrex Corporation
Campbell, California

C. M. Johnson
Brown and Root, Inc.
Houston, Texas

This document has been produced under the auspices and review of the Seismic Analysis Committee of the ASCE Nuclear Structures and Materials Committee.

SEISMIC ANALYSIS COMMITTEE MEMBERSHIP:

R.P. Kennedy, Chairman
A.H. Hadjian, Assistant Chairman

T. Blejwas	R. P. Kassawara
N. Chauhan	L. Malik
H. J. Dahlke	D. P. Moore
L. Gerdes	R. Murray
I. K. Ghosh	M. O'Rourke
E. C. Goodling	J. W. Reed
R. C. Guenzler	D. V. Reddy
A. K. Gupta	R. Sadigh
J. F. Halsey	P. D. Smith
A. Haldar	S. Singh
W. W. Hays	C. V. Subramanian
R. J. Hunt	R. Stuart
I. Idriss	W. A. VonRiesemann
J. J. Johnson	R. Weaton
D. D. Kana	P. Yanev
J. K. Khanna	

FLUID STRUCTURE INTERACTION DURING SEISMIC EXCITATION

TABLE OF CONTENTS

		Page
1.0	INTRODUCTION	1
2.0	OVERALL THEORETICAL DEVELOPMENT	2
	2.1 Hydrodynamic/Structural Formulation	2
	2.1.1 General Theory	2
	2.1.2 Application to Fluid Containers	4
	2.1.2.1 Liquid Response in a Rigid Container	5
	2.1.2.2 Elastic Container Response	7
	2.1.3 Application to Submerged Structures	10
	2.1.4 Application to Floating Structures	13
	2.2 Response Prediction Methods	16
	2.2.1 Time History	16
	2.2.2 Response Spectrum	16
	2.2.3 Power Spectral Density	18
3.0	LIQUID CONTAINERS	20
	3.1 Overview	20
	3.2 Typical Response to Seismic Excitation	20
	3.3 Typical Design Methods	22
	3.3.1 Continuum Transfer Function Approach	22
	3.3.2 Finite Element Formulations	24
	3.3.3 Simple Analytical Models	24
	3.4 Upright Cylindrical Container	26
	3.4.1 Modified Housner Method	27
	3.4.1.1 Sloshing Response in Shallow Tanks (H/R<1.5)	27
	3.4.1.2 Elastic Tank Horizontal Mode	30
	3.4.1.3 Horizontal Impulsive Response (H/R<1.5)	32
	3.4.1.4 Horizontal Sloshing and Impulsive Response in Tall Tanks (H/R>1.5)	32
	3.4.2 Vertical Impulsive Response	33
	3.4.3 Alternate Simple Analytical Models	34

TABLE OF CONTENTS - Cont'd.

		Page
3.4.4	Procedure for Elevated Tank	35
3.4.5	Consideration of Base Fixity and Buckling	37
3.5	Sloshing in Tanks of Various Geometries	39
3.5.1	Rectangular Tank	39
3.5.2	Horizontal Cylindrical Tank	39
3.5.3	Annular Cylindrical Tank	43
3.5.4	Toroidal Tank	45
3.5.5	Other Geometries	47
4.0 SUBMERGED STRUCTURES - ADDED MASS AND DAMPING		48
4.1	Overview	48
4.2	Single Isolated Members	48
4.2.1	Added Mass for Single Isolated Members	49
4.2.2	Added Damping for Single Isolated Members	53
4.3	Multiple Members	54
4.3.1	Hydrodynamic Coupling for Groups of Cylinders	54
4.3.2	Hydrodynamic Coupling for Rigid Members Surrounded by a Rigid Circular Cylinder	56
4.3.3	Hydrodynamic Coupling for Flexible Coaxial Cylinders	59
4.3.4	Damping for Multiple Members	59
5.0 FLOATING STRUCTURES		60
5.1	Overview	60
5.2	AGS Power Plant	60
5.3	Amplification of Water Transmitted Seismic Excitations	64
5.4	Basin Response	64
5.5	Fluid-Structure Interaction Analysis	65
5.5.1	Structure Idealization	65
5.5.2	Fluid Idealization	66
5.5.3	Structure-fluid Coupling	68
5.6	Mooring System	68
REFERENCES		69

TABLES

		Page
1.	Structures and Excitations of Concern	11
2.	Simple Model Equations for Cylindrical Tank	29
3.	Added Damping Values Projected in Figure 16 for Single, Isolated Structures	55

FIGURES

1.	Coordinate System for Seismic Excitation of Liquid Storage Tank	21
2.	Frequencies and Mode Shapes	21
3.	Modal Response of a Flexible Upright Circular Cylindrical Liquid Tank	23
4.	Dynamic Model of Tank with Fluid, Rigidly Supported on the Ground	25
5.	Simple Analytical Model for Base Fixed Flexible Cylindrical Tank	28
6.	Dynamic Models of Elevated Tanks	28
7.	Peak Liquid Responses in Model Storage Tank of Upright Cylinder	36
8a.	Elephant-foot Buckle in Wente "broad" Tank	38
8b.	Diamond-shaped Buckle in Wente "tall" Tank	38
9.	Coordinate System for Rectangular Tank	40
10.	Excitation Arrangement for Horizontal Cylindrical Tank	40
11.	Liquid Slosh Response for Simulated Seismic Excitation of Principal Axes in Horizontal Cylinder	42
12.	Circular Cylindrical Ring (Annular Tank)	44
13.	Seismic Slosh Response in Annular Tank Under El Centro, 1940 Earthquake	44
14.	Calculated and Experimental Values of the First Two Natural Frequencies for Horizontal Orientation of Toroid	46
15.	Seismic Slosh Response in Torus Tank Under Modified El Centro, 1940 Earthquake	46
16.	Submerged Body and Its Virtual Mass	50
17.	Circular Cylinder and Rectangular Plates	50
18.	Relative Effect of Virtual Mass in Parallelepipeds Square Side Moving Broadside On	52
19.	Liquid Mass Correction Factor in Circular Pier	52
20.	Amplitude/Diameter Value for Linear Damping Oscillating Submerged Circular Cylinders	57
21.	Two-Body Motion with Fluid Coupling	57
22.	Atlantic Generating Station	61
23.	Features of the Atlantic Generating Station	62
24.	Concept of a Tension Leg Platform	63

NOMENCLATURE

a,b	-	Dimensions of rectangular container or radii of concentric annular container or radii of toroidal container
c	-	Velocity of sound in fluid
C_M, C_D	-	Effective inertia and drag coefficients
C_x, C_y, C_z	-	Structural damping effective in x,y,z-directions for continuous formulation
D, L	-	Width and Length dimensions of submerged body
E	-	Elastic modulus
$\{F\}$	-	Generalized external force vector matrix
g	-	Acceleration of gravity
$G_i(\omega)$	-	Power spectral density of response at point i
$G_{x_0}(\omega)$	-	Power spectral density of excitation x_0
H	-	Vertical height of liquid in container
$[H], [T], [G]$	-	Mass, radiation damping, and stiffness matrices for discrete formulation of fluid
$H_{jx}(\omega)$	-	Transfer function between response at point j to excitation in x-direction
I	-	Cross-sectional moment of inertia
K_x, K_y, K_z	-	Structural stiffness effective in x,y,z-directions for continuous formulation
ℓ	-	Height of cylindrical container
L_{ij}	-	Linear operators for structures, Eqn. (2.1-8)
$[L]$	-	Fluid/structure interface matrix
M_x, M_y, M_z	-	Structural masses effective in x,y,z-directions for continuous formulation
$[M], [C], [K]$	-	Mass, damping, and stiffness matrices for discrete formulation of structure
p_{Bx}, p_{By}, p_{Bz}	-	Boundary pressure forces in x,y,z-directions

NOMENCLATURE - Cont'd.

p	-	Liquid pressure
p_o	-	Ullage pressure
q_{ni}	-	Generalized displacement coordinate for mode n in i-direction
r,θ,z	-	Cylindrical coordinates
R	-	Radius of cylindrical container, or cross-sectional radius of toroidal container
$\{R\}$	-	Load vector matrix
$S_d(\omega), S_v(\omega), S_a(\omega)$	-	Displacement, velocity, and acceleration response spectrum
t_w	-	Wall thickness of container
w,v,u	-	Structural elastic displacements in x,y,z-directions
$W_o, W_1, M_o, M_1, h_1, h_o, P_o$, etc.		Parameters for modified Housner slosh model (see Table 2, and Figures 4,5,6).
x,y,z	-	Cartesian coordinates with z-vertical
$\ddot{X}, \ddot{Y}, \ddot{Z}$	-	Seismic accelerations in x,y,z-directions
ZPA	-	Zero period acceleration
β_{ni}	-	Damping ratio for normal mode n in i-direction
Γ_{ni}	-	Modal participation factor for mode n in i-direction
η	-	Liquid wave height at free surface
η^*	-	Peak response value for liquid wave height
λ	-	Liquid bulk modulus
ν	-	Poisson's ratio
ξ_r	-	Eigenvalue for rth mode
ρ	-	Liquid mass density
ρ_s	-	Shell (container) density

NOMENCLATURE - Cont'd.

σ_i	-	RMS value of a response at point i
σ_θ	-	Circumferential stress
ϕ	-	Velocity potential for liquid
ϕ_{nc}	-	Liquid velocity potential normal modal function for mode n in nonmoving, elastically vibrating container
ϕ_{ns}	-	Liquid velocity potential normal mode function for slosh in mode n in nonmoving container
ϕ_r	-	Liquid velocity potential resulting from seismic motion of container
χ_n	-	Liquid surface slosh normal modal function
ψ_{ni}	-	Structural elastic displacement normal modal function in mode n for i-direction
ω_{ni}	-	Natural frequency for mode n in i-direction

1.0 INTRODUCTION

Hydrodynamic loading and other fluid/structure interaction effects need to be considered in the design of structures which contain, surround, or are submerged in fluids when subject to seismic excitation. Nuclear reactor facilities typically include numerous fluid containers of a variety of geometries and sizes, and in many cases substructures are submerged within the contained fluids. Therefore, a workable approach must be used in the design of these facilities.

The objective of this document is to provide a guide for the design of liquid containers, submerged structures, and floating structures. However, in each case, only a general approach to each area is presented with appropriate references that can be consulted for more details. In many cases, detailed theoretical analyses of the problem are available, and literature sources for that information will be cited. However, for some aspects of the problem, complete solutions are not yet included within the state-of-the-art. These areas will also be identified. Although complex solutions are available for some applications, the emphasis herein is on simpler, design methods. We begin with only a brief overview of the analytical background, in order to indicate the framework on which the design methods are based.

2.0 OVERALL THEORETICAL DEVELOPMENT

2.1 Hydrodynamic/Structural Formulation

2.1.1 General Theory

Neglecting fluid viscosity and assuming small displacements, the motion of a fluid is governed by the following differential equation [32]:

$$c^2 \nabla^2 \phi = \frac{\partial^2 \phi}{\partial t^2} \qquad (2.1\text{-}1)$$

where ϕ = a velocity potential such that

$$\frac{\partial \delta_i}{\partial t} = + \frac{\partial \phi}{\partial x_i} \qquad (2.1\text{-}2)$$

δ_i = liquid displacement along the Cartesian coordinates (x-horizontal, y-horizontal, z-vertical)
c = velocity of sound = $\sqrt{\lambda/\rho}$
λ = Bulk Modulus
ρ = mass density
∇^2 = Laplace's operator, and
t = time.

For liquid responses in the seismic frequency range (i.e., $\leq$ 33 Hz), the fluid may be assumed to be incompressible, Equation (2.1-1) becomes

$$\nabla^2 \phi = 0 \qquad (2.1\text{-}3)$$

Along with this case, the fluid pressure p is governed by the linearized Bernoulli equation

$$\frac{\partial \phi}{\partial t} + \frac{1}{\rho} p + g z = 0 \qquad (2.1\text{-}4)$$

Furthermore, the velocity potential ϕ can be separated into two parts

$$\phi = \phi_1 + \phi_2 \qquad (2.1\text{-}5)$$

where ϕ_1 is the velocity potential of the fluid for a fixed coordinate system, and ϕ_2 is the velocity potential which allows for a moving coordinate system.

Solution of Equations (2.1-3) and (2.1-5), with associated pressure given by Equation (2.1-4) must be obtained with the appropriate boundary conditions of the problem. At a free surface the boundary condition requires that no fluid particles leave the surface (i.e., the vertical surface velocity equals the vertical fluid particle velocity), and the ullage (gas over liquid) pressure is a constant. For these conditions the free surface boundary condition reduces to the waveheight η satisfying the following:

$$\frac{\partial \eta}{\partial t} - \frac{\partial \phi}{\partial z} = 0 \qquad (2.1\text{-}6a)$$

at free surface

$$\frac{\partial \phi}{\partial t} + g\eta + \frac{p_o}{\rho} = 0 \qquad (2.1\text{-}6b)$$

Furthermore, at any boundary of the structure the normal fluid velocity must equal the normal structure velocity. This results in

$$\frac{\partial \phi}{\partial n} = v_n \qquad (2.1\text{-}7)$$

where v_n is the normal velocity of the structure.

Additional governing equations for elastic deformation of the structure must also be included. If the structural elastic displacements are w, v, u relative to a Cartesian coordinate system x, y, z, which is moving with accelerations $\ddot{X}$, $\ddot{Y}$, $\ddot{Z}$, the equations often may be simplified to the form:

$$L_{11}(w) + L_{12}(v) + L_{13}(u) + C_x \dot{w} = -M_x(\ddot{w} + \ddot{X}) + P_{Bx} \qquad (2.1\text{-}8a)$$

$$L_{21}(w) + L_{22}(v) + L_{23}(u) + C_y \dot{v} = -M_y(\ddot{v} + \ddot{Y}) + P_{By} \qquad (2.1\text{-}8b)$$

$$L_{31}(w) + L_{32}(v) + L_{33}(u) + C_z \dot{u} = -M_z(\ddot{u} + \ddot{Z}) + P_{Bz} \qquad (2.1\text{-}8c)$$

where L_{ij} are linear operators which form the coupled stiffness of the structure. The structural damping coefficients are presumed to be of such a form that their decoupling can be achieved, which is consistent with usual engineering practice. Furthermore, the structural masses are shown as decoupled, which may not always be the case. Likewise, some degree of coupling may also occur through the liquid pressures P_{Bx}, P_{By}, and P_{Bz}, which are obtained from Equation (2.1-4) and imposing the boundary conditions at the tank bottom and wall, and at the liquid surface. In the latter event, a more complex formulation is required, details of which can be obtained from several of the references given herein.

It should be recognized that all of the above equations have been written as if both the fluid and the structure are continuous systems. However, corresponding matrix equations can be written with a discrete system formulation as well. The structural equations (2.1-8) are almost identical to that for a structure alone, except that the additional fluid loading term is present.

The essence of the fluid/structure interaction problem lies in simultaneous solution of the above-described equations, consistent with the boundary conditions present. The approach appropriate for this combined fluid/structure interaction solution depends very much on the exact type of the three problem areas identified in the INTRODUCTION above, and also on the geometry present. In many cases the fluid response can be effectively decoupled from the structural response to some degree or another. In other cases the effect of the fluid loading can be approximated simply as a mass or damping added to the structure. In all cases, the solution involves a two part approach which includes some form of separation of space variables so that there results a set of differential equations in time.

2.1.2 Application to Fluid Containers

Solution of the above problem for the design of fluid containers has been studied in considerable depth by Housner [1,2] as well as others [34,50], and similar solutions have been pursued for aerospace applications [3,4]. The immediate applicability of most of the latter data to the seismic design problem is summarized in an extensive review paper by Kana [5].

The seismic response problem for typical containers can be separated into a decoupled set of equations which govern the slosh modes of the liquid, and a coupled set which govern structural modes that are influenced by mass-like pressure loading of the liquid. The boundary conditions of the specific problem must be considered for the solution of these equations. A brief summary will be given for this approach in terms of the normal modes of the system. Because of the frequency separation of typical liquid surface slosh and elastic container wall responses, the approach is given in terms of two separate problems.

2.1.2.1 Liquid Response in a Rigid Container

Liquid surface slosh occurs at such low frequencies relative to elastic wall response that consideration of liquid response in a rigid container is appropriate for prediction of this part of the problem. As indicated in Eqn. (2.1-5), the liquid potential can be separated into two parts:

$$\phi(x,y,z,t) = \phi_s(x,y,z,t) + \phi_r(x,y,z,t) \qquad (2.1-9)$$

where $\phi_s(x,y,z,t)$ is a slosh potential developed from the potential for liquid motion in a non-moving container, and $\phi_r(x,y,z,t)$ is a potential which allows for rigid body motion of the container (moving coordinate system).

For this case the liquid surface response η can be expanded into a series of normal modes:

$$\eta_i(x,y,t) = \sum_{n=1}^{\infty} \chi_{ni}(x,y)\, q_{ni}(t) \qquad (2.1-10)$$

where $i = x, y,$ or z, and $\chi_{ni}(x,y)$ are normal modal functions which are related to the velocity potential. This potential can also be expanded as

$$\phi_i(x,y,z,t) = \left[\sum_{n=1}^{\infty} \phi_{ns}(x,y,z)\, q_{ns}(t) + \phi_r(x,y,z)\, q_r(t)\right]_i \qquad (2.1-11)$$

where $\phi_{ns}(x,y,z)$ are normal modal functions which satisfy the fluid equation and appropriate boundary conditions in a nonmoving container. For this case the elastic wall response is

$$u,v,w = 0$$

although the container moves with accelerations $\ddot{X}, \ddot{Y}, \ddot{Z}$.

Evaluation of the free surface boundary condition Eqn. (2.1-6a) requires that

$$\dot{q}_{ni}(t) = \left[q_{ns}(t)\right]_i \qquad (2.1\text{-}12a)$$

$$\frac{\partial}{\partial z} \phi_{ns}(x,y,z) \bigg|_{s,i} = \chi_{ni}(x,y) \qquad (2.1\text{-}12b)$$

$$\frac{\partial}{\partial z} \phi_r(x,y,z) \bigg|_{s,i} = 0 \qquad (2.1\text{-}12c)$$

Furthermore, use of these results, the boundary condition Eqn. (2.1-7), and assumption of a damping term in the boundary condition Eqn. (2.1-6b) along with use of the orthogonality properties of the modal functions results in equations of the form

$$\ddot{q}_{ni} + 2\beta_{ni}\omega_{ni}\dot{q}_{ni} + \omega_{ni}^2 q_{ni} = -(\ddot{X},\ddot{Y},\ddot{Z}) \, \Gamma_{ni} \qquad (2.1\text{-}13)$$

where the modal participation factor is

$$\Gamma_{ni} = \frac{\left[\rho \int_s \phi_r(x,y,z_s)\phi_{ns}(x,y,z_s) \, dx \, dy\right]}{\left[\rho \int_s \phi_{ns}^2(x,y,z_s) \, dx \, dy\right]_i} i \qquad (2.1\text{-}14)$$

where s denotes evaluation at the free surface. It is understood that three equations corresponding to $i = x,y,z$ are included. However, linearized liquid surface slosh is known to occur only for horizontal (x or y) excitation, and significant response occurs only for the lowest several modes. For z-excitation (vertical), $\Gamma_{nz} = 0$, so that no liquid surface slosh occurs.

As indicated in Equation (2.1-9), the total liquid potential includes an additional part $\phi_r(x,y,z,t)$ which results from rigid body motion of the container. Further investigation of this part of the potential for response parallel to the excitation axis leads to a non-resonant part of the liquid which produces an inertial (impulsive) load on the lower part of the container. The effective rigid mass of the liquid at high frequencies is equal to the difference of the total liquid mass and the sum of the sloshing masses, and like the sloshing masses, may be different for each of the three orthogonal excitation axes.

2.1.2.2 Elastic Container Response (Cylindrical Geometry)

Response of the elastic container typically occurs at frequencies well above those for significant liquid surface slosh. Therefore, the dominant effect of the fluid in this case takes the form of an inertial loading on the container. However, it is difficult to discuss the solution of the elastic tank response without specifying the tank geometry. Therefore for simplicity of illustration, we consider an upright cylindrical tank of radius R.

The elastic wall responses can be expanded into a series of normal modes

$$w = \sum_{n=1}^{\infty} \psi_{nr}(\theta,z) \, q_{nr}(t) \qquad (2.1\text{-}15a)$$

$$v = \sum_{n=1}^{\infty} \psi_{n\theta}(\theta,z) \, q_{n\theta}(t) \qquad (2.1\text{-}15b)$$

$$u = \sum_{n=1}^{\infty} \psi_{nz}(\theta,z) \, q_{nz}(t) \qquad (2.1\text{-}15c)$$

where ψ_{nr}, $\psi_{n\theta}$, ψ_{nz} represent container normal modal functions which satisfy Eqns. (2.1-8) when $\ddot{X} = \ddot{Y} = \ddot{Z} = 0$. Associated with this are liquid pressure responses for which surface slosh motions are negligible. Thus with g = 0, Eqn. (2.1-4) becomes

$$p = -\rho \frac{\partial \phi}{\partial t} \qquad (2.1\text{-}16)$$

However, $\phi(r,\theta,z,t)$ still must satisfy all the other conditions of Section 2.1.1. It can be expanded as

$$\phi(r,\theta,z,t) = \sum_{n=1}^{\infty} \phi_{nc}(r,\theta,z) \, q_{nc}(t) + \phi_r(x,y,z) \, q_r(t) \qquad (2.1\text{-}17)$$

where $\phi_{nc}(r,\theta,z)$ are normal modal functions which satisfy the fluid equations with the assumption of $g = 0$, and the boundary conditions for a nonmoving elastic container with vibrating walls. Furthermore, $\phi_r(r,\theta,z) \, q_r(t)$ is a velocity potential which satisfies the fluid equations, and the boundary conditions of a rigid container moving with the prescribed base accelerations $\ddot{X}, \ddot{Y}, \ddot{Z}$. Note that because of large frequency separation, sloshing is not considered for this part of the problem, and therefore $\phi_r(r,\theta,z) \, q_r(t)$ is the same rigid body portion of the fluid potential which represents the impulsive mass difference of the total fluid mass and the sum of the sloshing fluid masses, as was discussed in Section 2.1.2.1.

First, consider container motion in one horizontal (x-direction) only. From Eqns. (2.1-7), (2.1-15a) and 2.1-17 there results at the container wall ($r = R$):

$$\frac{\partial \phi}{\partial r}\bigg|_R = \sum_{n=1}^{\infty} \frac{\partial}{\partial r}[\phi_{nc}(r,\theta,z)] \, q_{nc}(t)\bigg|_R$$

$$+ \frac{\partial}{\partial r}[\phi_r(r,\theta,z)] \, q_r(t)\bigg|_R = \frac{\partial w}{\partial t} + \dot{X} \cos \theta \qquad (2.1\text{-}18)$$

where $\dot{X}$ is the velocity of the base motion along the x-axis. But, from Eqn. (2.1-15a) there results:

$$\frac{\partial w}{\partial t} = \sum_{n=1}^{\infty} \psi_{nr}(\theta,z) \, \dot{q}_{nr}(t) \qquad (2.1\text{-}19)$$

Therefore to satisfy Eqn. (2.1-18) it is necessary that

$$\frac{\partial}{\partial r}[\phi_{nc}(r,\theta,z)]_R = \psi_{nr}(\theta,z) \qquad (2.1\text{-}20a)$$

$$\frac{\partial}{\partial r}[\phi_r(r,\theta,z)]_R = \cos \theta \qquad (2.1\text{-}20b)$$

and that
$$q_{nc}(t) = \dot{q}_{nr}(t) \tag{2.1-20c}$$
$$q_r(t) = \dot{X} \tag{2.1-20d}$$

Therefore for acceleration $\ddot{X}$ of the tank base, at the wall boundary the pressure can be approximated by

$$p_R = -\rho R \sum_{n=1}^{\infty} \psi_{nr}(\theta,z) \ddot{q}_{nr}(t) - \rho R \cos\theta \ddot{X} \tag{2.1-21}$$

This expression along with $P_{By} = P_{Bz} = 0$ are now substituted into Eqns. (2.1-8a), which must be written for cylindrical shell geometry. However, use of the normal modal functions (2.1-15) still results in a coupled set of three equations. Details of solutions for the coupled equations with horizontal only excitation have been given by Haroun [73]. However, Veletsos [50] and others have developed decoupled equations either by elimination of variables, or by ignoring relatively weakly coupled terms. For such cases, the governing equation for the upright cylinder with horizontal only excitation can be reduced to the form

$$\ddot{q}_{nr} + 2\beta_{nr}\omega_{nr}\dot{q}_{nr} + \omega_{nr}^2 q_{nr} = -\ddot{X} \left| \frac{m_{cr} + m_{\ell r}}{m_{cw} + m_{\ell w}} \right| \tag{2.1-22}$$

where the respective masses of the modal participation factor are given for the container by:

$$m_{cr} = \int_C M(\theta,z) \psi_{nr}(\theta,z) \, d\theta \, dz \tag{2.1-23a}$$

$$m_{cw} = \int_C M(\theta,z) \psi_{nr}^2(\theta,z) \, d\theta \, dz \tag{2.1-23b}$$

and for the liquid by

$$m_{\ell r} = \rho R \int_R \cos\theta \, \psi_{nr}(\theta,z) \, d\theta \, dz \tag{2.1-23c}$$

$$m_{\ell w} = \rho R \int_R \psi_{nr}^2(\theta,z) \, d\theta \, dz \tag{2.1-23d}$$

A similar equation can be developed for the vertical motion of the system.

2.1.3 Application to Submerged Structures

An extensive review of fluid/structure interaction for submerged structures has been given by Dong [6]. Table (1), which is taken from that source, lists some typical structures, loads, and parameters for this problem. Here not only seismic, but other transient pressure loads must be considered. Typically, the hydrodynamic loading for structures in this case is approximated by mass and/or damping terms. That is, for x-direction motion the hydrodynamic loading p_{Bx} which appears in Eqn. (2.1-8a), is estimated by Morison's equation:

$$p_{Bx} = 1/2 \, \rho \, C_D \, |\dot{w}_r| \dot{w}_r + \rho C_m \ddot{w}_r \qquad (2.1\text{-}24)$$

where $\dot{w}_r$ is the relative velocity between the structure and the water. C_D is a drag or damping coefficient, and C_M is an effective inertia coefficient. The load given by Eqn. (2.1-24) is expressed per unit area normal to the direction of the motion. If the structure moves through otherwise still water, C_M is called the added or virtual mass coefficient and represents the volume of liquid affected, or following, the structural motion. If the structure is stationary and the liquid is in motion, the volume of liquid displaced by the structure is added to the virtual mass coefficient.

Equation (2.1-24) implicitly incorporates a number of assumptions and restrictions. The motion of the fluid or the structure is assumed to be slow enough that acoustic effects, e.g., radiation damping, are negligible; this is not a particularly restrictive assumption for the applications of interest here. The structure is assumed to move transversely as a rigid body, and must be of long slender geometry. Most importantly, the damping forces are assumed to be separable from the inertial forces. To make things complicated, the coefficients C_M and C_D are functions not only of the geometry of the structure but also, in many cases, of fluid properties and of the motion. For example, when the structure vibrates, C_M

TABLE 1. STRUCTURES AND EXCITATIONS OF CONCERN
(From Reference 6)

Structures	Excitations
Spent-fuel storage racks	Seismic
Main steam-relief valve line	Pressure relief Blowdown-induced loads Seismic
Internals of reactor vessel	Blowdown-induced loads Seismic

REPRESENTATIVE SIZES AND NATURAL FREQUENCIES OF STRUCTURES OF CONCERN

Structure	Size	Natural frequency	Condition
Fuel elements	$\sim$ 0.5 in. D		
Fuel bundles, BWR	$\sim$ 5.5 x 5.5 in.	$\sim$3 Hz	In water
Fuel bundles, PWR	$\sim$10 x 10 in.	$\sim$3 Hz	In water
Fuel racks: firm (1)		17 to 33 Hz	Full and in water
Fuel racks: firm (2)		10 to 20 Hz	Full and in water
Fuel racks: firm (3)		6 to 9 Hz	Full and in water
Fuel racks: firm (4)		$\sim$ 12 Hz	Full in air
Fuel racks: firm (4)		$\sim$ 10.5 Hz	Full in water
Fuel racks: firm (5)		$\sim$ 1.15 Hz	Full in water
Main steam-relief valve line	8 in. D, 72 in. L	0.5 Hz	In air
	8 in. D, 72 in. L	1.2 Hz	In air
	8 in. D, 396 in. L	0.02 Hz	In air
	8 in. D, 396 in. L	0.04 Hz	In air
	12 in. D, 72 in. L	0.8 Hz	In air
	12 in. D, 72 in. L	1.8 Hz	In air
	12 in. D, 396 in. L	0.03 Hz	In air
	12 in. D, 396 in. L	0.06 Hz	In air
Reactor core barrel		40 Hz	In air
		10 Hz	In water

depends on the amplitude and frequency of the vibration, as does C_D. For a linearly accelerated motion (i.e., $\dot{w}_r = \ddot{w}_r t$), C_D is a function of the acceleration and the time since the motion began, being close to zero near the start when the flow around the structure resembles an inviscid, potential flow, and taking on the steady flow value after the boundary layer develops. When other fixed or moving structures are nearby, C_D and especially C_M are strong functions of the geometry and proximity of the other structures. C_M and C_D also depend on whether the structure is near to or penetrates the liquid free surface; if so, C_M is generally decreased and C_D increased in comparison to similar cases for deeply submerged structures. Because of this complexity, the available surveys emphasize data tabulations of C_M and C_D for representative structures as a function of representative conditions. Dong [6] has concluded that the available experimental data and design/computing techniques for C_M and C_D are insufficient and probably inaccurate for main steam-relief valve lines, reactor vessel internals, and spent fuel racks. Most designs are therefore based on what is considered to be conservative engineering judgment.

When only small amplitude motions are considered, the velocity squared drag-like term in Eqn. (2.1-24) can be replaced by a linear velocity damping term of the form $C\dot{w}_r$; C can, however, be a function of $\dot{w}_r$ and $\ddot{w}_r$, or alternatively, of ω and ω^2 where ω is the vibration frequency. Equation (2.1-24) can also be modified to allow structural flexibility by using various approximate theories, such as two-dimensional "strip" theory, to compute C_M and C_D distributions along the structural axis. (This approximation is most applicable to bending or beam-like motions of a slender structure. Various modifications can be made to incorporate shell-like vibration modes.) In this case, $\dot{w}$ and $\ddot{w}$ are the relative velocity and acceleration at the structural location in question. Since the geometry of typical underwater structures and the boundaries of the liquid are often rather complex, a discrete formulation of the problem is convenient. Therefore, for a discrete system formulation, the structural response can be expressed in terms of normal modes, $\{w\} = [\psi]_x \{q\}_x$, and the loading can be expressed in matrix form as:

$$\{p_{Bx}\} = [C_D][\psi]_x \{\dot{q}\}_x + [C_M][\psi]_x \{\ddot{q}\}_x \qquad (2.1\text{-}25)$$

Putting this loading into the equation of motion for the structure (i.e., Eqn. (2.1-8a) then gives:

$$\lfloor I \rfloor \{\ddot{q}\}_x + \lfloor 2\beta_{nx}\omega_{nx} \rfloor \{\dot{q}\}_x$$
$$+ \lfloor \omega_{nx}^2 \rfloor \{q\}_x = [\psi]_x^T \{F\} \qquad (2.1\text{-}26)$$

Here $\lfloor I \rfloor$ is the unity matrix which results from normalization by a combined structural mass and added fluid mass matrix, $\lfloor 2\beta_{nx}\omega_{nx} \rfloor$ the coupled fluid/structural modal damping, and $\{F\}$ the generalized external-forcing matrix, excluding any forcing due to the liquid. The coupled set of Eqns. (2.1-26) can be reduced to an uncoupled set of equations similar to Eqn. (2.1-22) by various approximations. Therefore, the same methods of solutions apply thereafter.

2.1.4 Application to Floating Structures

There has been some interest in offshore siting of nuclear power plants because of the relatively lower interference by other community activities for such a location. A comprehensive review for the design and analysis of floating nuclear power plants has been presented by Thangam Babu and Reddy [54], as well as others. The topics covered include offshore concept evaluation (as discussed by several investigators), siting, environmental and design considerations, fabrication and installation, static and dynamic analyses, and model studies. Other topics discussed were fatigue and crack propagation, stability criteria, mooring foundation analysis, safety, waste disposal, instrumentation, corrosion control, noise and vibration levels, legal aspects and cost estimates.

The coverage in this section will be restricted to fluid-structure interaction analysis. Reviews of earlier work, new formulations, and applications have been presented by Thangam Babu and Reddy [55, 56, 57], Arockiasamy et al. [58], Reddy et al. [59], Thangam Babu, Arockiasamy, and Reddy [60], Thangam Babu, Reddy, and Arockiasamy [61], Arockiasamy, Thangam Babu, and Reddy [62], and Thangam Babu [63].

As indicated in Section 2.1.1 the coupled fluid-structure interaction problem can be dealt with different degrees of simplifications and complexities. Neglecting the nonlinearities, in its simplest form the vibration of a floating body in an incompressible and inviscid fluid of finite depth with a free wave surface can be described by LaPlace's equation with specified boundary conditions at the free surface, the sea bottom, and the fluid-structure interface, together with radiation damping simulating the infinite medium. Classical techniques and closed form solutions to this type of problem have been restricted to well-defined basin and structure geometries and rigid floating bodies. Complex offshore facilities and basin geometries render these methods undesirable. The finite element method becomes handy in such cases, where the fluid depth is finite and the basin and structure geometries are of arbitrary shape; the flexibility of the structure can be taken into account with little difficulty. A discrete system theoretical formulation yields the following coupled unsymmetric matrix equation.

$$\begin{bmatrix} M & 0 \\ \rho L^T & H \end{bmatrix} \begin{Bmatrix} \ddot{w} \\ \ddot{p} \end{Bmatrix} + \begin{bmatrix} C & 0 \\ 0 & T \end{bmatrix} \begin{Bmatrix} \dot{w} \\ \dot{p} \end{Bmatrix} + \begin{bmatrix} K & L \\ 0 & G \end{bmatrix} \begin{Bmatrix} w \\ p \end{Bmatrix} = \begin{Bmatrix} R \\ 0 \end{Bmatrix} \quad (2.1\text{-}27)$$

where

[M], [C] and [K] = mass, damping, and stiffness matrices of the structure,

[H], [T] and [G] = mass, radiation damping, and stiffness matrices of the fluid,

[L] and [L]^T = fluid-structure interface matrix and its transpose,

w, $\dot{w}$, and $\ddot{w}$ = displacement, velocity, and acceleration vectors of the structure,

p, $\dot{p}$, and $\ddot{p}$ = pressure, its first, and second derivatives

ρ = fluid mass density, and

R = load vector.

Equation (2.1-27) represents a set of discrete equations analogous to Eqns. (2.1-8) combined with several appropriate fluid equations in terms of the pressure p. Although the individual matrices in Equation (2.1-27) are symmetric, (with the exception of the interface matrix), the coupled equation is unsymmetric and unbanded. It has to be converted into two symmetric equations to apply the standard numerical integration technique. In a time-step integration as presented by Thangam Babu and Reddy [56] and Thangam Babu [63], the unsymmetric coupled equation is transformed into two symmetric equations. Without making any simplifications or assumptions, but by making use of the unique nature of the matrix coupling the fluid and structural matrices, L, the independent equations are manipulated to yield two symmetric matrices. The Newmark-β and the Wilson-θ numerical interpretation schemes have been used to solve these equations.

2.2 Response Prediction Methods

The development in Sections 2.1.2 through 2.1.4 shows that the fluid/structure interaction problem can be reduced to a coupled set of differential equations in time. At this point the timewise solution methods that are applicable are generally similar to those used in dynamic analysis of structures alone. However, we will summarize further at this point, to emphasize the utility of much design data that are available from earlier aerospace research [3,4]. The details are given by Kana [5].

2.2.1 Time History

Generally, time history methods involve direct integration of Equations (2.1-3) and (2.1-8) either by analog computer or by digital step integration schemes. Some details of available methods have been summarized by Healey, et al [7]. If possible, a normal mode approach using Equations (2.1-13) and 2.1-22) is preferred. However, if nonlinearities are present, or if damping cannot be considered proportional to mass or stiffness, then Equations (2.1-3) and (2.1-8) must be dealt with directly. In either case, the approach is more complex than the spectral methods described in the next paragraph.

2.2.2 Response Spectrum

Earthquake response spectrum methods can be applied to any system for which a normal mode solution is sought. Here we will include some detailed development in order to emphasize how transfer functions for liquid response derived in earlier work [3,4] can be used in earthquake response prediction of liquid slosh by this method. The development is a condensation of that given by Kana [5]. In Eqn. (2.1-13) we focus our attention on the response of one (nth) mode at a time. The equation of motion for x-excitation (horizontal) is:

$$\ddot{q}_{nx} + 2\beta_{nx}\omega_{nx}\dot{q}_{nx} + \omega_{nx}^2 q_{nx} = -\Gamma_{nx}\ddot{X} \qquad (2.2\text{-}1)$$

where Γ_{nx} is given by Eqn. (2.1-14). Furthermore, from Eqn. (2.1-10) the nth mode eigenvector $\chi_{nx}(x,y)$ is related to the nth mode physical displacements at point (x_j, y_j) by

$$\eta_n(x_j, y_j) = \chi_{nx}(x_j, y_j) q_{nx}(t) \tag{2.2-2}$$

The peak response q_{nx}^* of the nth mode to a seismic excitation can now be written in terms of the displacement response spectrum (i.e., analogous to page 554 of Clough and Penzien [8]) as

$$q_{nx}^* = \Gamma_{nx} S_{dx}(\omega_n) \tag{2.2-3}$$

where $S_{dx}(\omega_n)$ is the value of the base motion displacement response spectrum at frequency ω_n. Thus, by concentrating on the peak displacement $\eta_n^*(x_j, y_j)$ of the (x_j, y_j) node point in the wave mode shape, from Eqns. (2.2-2) and (2.2-3) one can write

$$\eta_n^*(x_j, y_j) = \chi_{nx}(x_j, y_j) \Gamma_{nx} S_{dx}(\omega_n) \tag{2.2-4}$$

By further development, Kana [5] has shown that the response spectrum coefficient can be expressed in terms of a transfer function as

$$\chi_{nx}(x_j, y_j) \Gamma_{nx} = 2\beta_n |H_{jx}(\omega_n)| \tag{2.2-5}$$

where $H_{jx}(\omega_n)$ is the value of the transfer function between response point (x_j, y_j) and the excitation at the resonance frequency for mode n. Thus, Eqn. (2.2-4) can be rewritten as

$$\eta_n^*(x_j, y_j) = 2\beta_n |H_{jx}(\omega_n)| S_{dx}(\omega_n) \tag{2.2-6}$$

Although this equation is written in terms of the slosh wave height η, a similar development can be based on the structural Eqns. (2.1-22) as well. Thus, $H_{jx}(\omega_n)$ can be the ordinary harmonic transfer function value for any response parameter of an arbitrary base excited structure. Furthermore, the equation can be used for prediction when the transfer function is obtained by analysis or by experiment. If response in multiple modes occurs, then the total response can be obtained by the usual square root of the sum of the squares (SRSS) [9], or some other combination. This result allows the direct use of much previous design data for prediction of earthquake response, as will be demonstrated in sections hereafter.

2.2.3 Power Spectral Density

The stochastic nature of earthquakes has always been recognized, and analytical methods for prediction of response under this concept have been given by Clough and Penzien [8], as well as others. However, formulation by random process theory has generally not been very popular compared to response spectrum methods, because of the inherent nonstationary random character of earthquakes. On the other hand, time average parameters for describing nonstationary processes have been defined by Bendat and Piersol [10] and furthermore, evidence has now been developed which indicates that typical seismic excitation can be considered essentially stationary during the strong motion portion of the time history, which is the most important for design purposes. Singh [11] has used this approach for prediction of structural response. Furthermore, Kaul [12] has shown that with this assumption, a direct transformation process from response spectrum to power spectrum, and vice versa is possible. Unruh and Kana [13] have developed a digital computer program for the transformation process. The latter technique is useful for development of dynamics formulations that are mathematically more rigorous by using random process theory rather than response spectrum relationships directly. The transformation from response spectrum to power spectrum, manipulation of power spectra to determine response, and subsequent retransformation back to response spectra can be accomplished without ever developing accompanying time histories.

In terms of the fluid structure seismic response problem, the stochastic response prediction equation takes the form

$$G_i(\omega) = |H_{ix_0}(\omega)|^2 G_{x_0}(\omega) \qquad (2.2\text{-}7)$$

This equation is analogous to Eqn. (2.2-6). Here $G_i(\omega)$ is the power spectral density of the response at some point i, $H_{ix_0}(\omega)$ is the transfer function between the response at i and the excitation at x_0 as before, and $G_{x_0}(\omega)$ is the power spectral density of the seismic excitation. If the power spectra are considered during the strong motion portion of the motion, then they are essentially stationary. If they are considered over the entire seismic event, then they may be averaged over time, and Eqn. (2.2-7) still remains valid.

It is further useful to recognize that the rms value (standard deviation) of the response can be obtained from

$$\sigma_i = \left[\int_0^{\omega_{max}} G_i(\omega) \, d\omega \right]^{1/2} \qquad (2.2\text{-}8)$$

Then, the peak value of response can be obtained as a statistical factor times σ_i. For the strong motion stationary portion of the time history, typically the peak value of a response η_i^* is

$$\eta_i^* \approx 3.0 \, \sigma_i \qquad (2.2\text{-}9)$$

3.0 LIQUID CONTAINERS

3.1 Overview

In this section, we consider more specifically the design of liquid containers. Inherently, this requires the prediction of motion responses, and thereby stress responses in the containers as a result of the earthquake excitation. More specific methods will follow from the discussion set forth in Section 2.0. Since liquid containers of a variety of geometries and sizes are used in nuclear plant facilities, some consideration of this will be given. However, since by far most of the data available to date has been developed for the upright cylinder, this particular case will be covered in most detail. Information for other geometries will be given, when available. We begin with a more general discussion of the upright cylinder, but recognize that the concepts apply to other geometries as well.

3.2 Typical Response to Seismic Excitation

A typical diagram of coordinates for prediction of liquid slosh in an upright cylindrical tank is given in Figure 1. Liquid slosh modes occur generally at very low frequencies (i.e., less than 0.5 Hz for typical container sizes). Examples for some frequencies for a typical size tank and a 1/30-scale model are given in Figure 2. The influence of the liquid is effectively separated into two parts, providing that the liquid modes and lowest structural modes are sufficiently separated. The upper portion of the liquid is influenced by the slosh motion, which is excited principally in antisymmetric modes by the horizontal ground motion. The lower part of the liquid acts like a rigid mass, which is excited by both horizontal and vertical ground motion. The net effect is to separate the total liquid mass into the sloshing or convective mass and the rigid or impulsive mass. The magnitude of each and their effective center of mass depends very much on the liquid depth. For vertical excitation, no slosh is excited, but impulsive mass effects are.

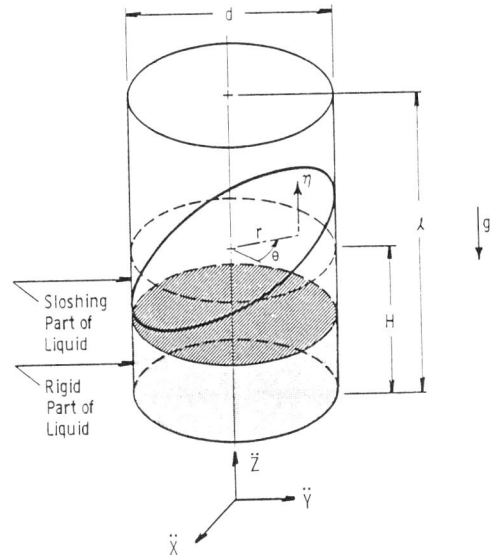

Figure 1. Coordinate System for Seismic Excitation of Liquid Storage Tank

Prototype Dimensions:
R_o = 60 ft (18.29 m)
H = 30 ft (9.14 m)

n	Prototype Freq., Hz	1/30 Model Freq., Hz	Mode Shape
1	0.135	0.738	
2	0.268	1.467	
3	0.339	1.856	
4	0.397	2.174	

Figure 2. Frequencies and Mode Shapes (From Reference 23)

Corresponding elastic shell responses of the tank are shown in Figure 3. For typical tanks, these modes occur above 2 Hz. Circumferentially these responses are proportional to cos $n\theta$, with θ being the circumferential coordinate. Breathing response ($n = 0$) corresponds to a closed-open organ pipe mode which is symmetrical about the vertical axis. This mode is excited by vertical excitation only, and produces hoop stress in the tank. Bending response ($n = 1$) is excited in medium to tall tanks by horizontal excitation. The frequency for this mode is very much influenced by the impulsive or rigid mass of the liquid. Finally, the shell modes ($n > 2$) occur in great numbers. However, the exact influence of these modes on the design stresses has not yet been established, although their effects have most often been neglected. In fact, the mechanism through which these modes are excited has not yet been completely established. Niwa[14] and Housner and Haroun [15,16] have agreed that such shell responses occur because of imperfections, or out of roundness of the cylinder. In any event, since the influence of such responses at present remains unestablished, we will not consider them further in these guidelines. Nevertheless, this should not be construed as their having negligible influence on buckling and other forms of failure.

3.3 Typical Design Methods

3.3.1 Continuum Transfer Function Approach

Because of the frequency separation of liquid slosh and elastic shell modes, determination of liquid response due to surface slosh only can usually be determined by the assumption of a rigid tank for this part of the problem. References [3] and [4] contain continuum solutions for harmonic excitation of liquids in rigid tanks of many different geometries. The solutions are developed in terms of various spacewise functions that are eigensolutions of Eqn. (2.1-3) in the given geometry. The results are summarized in terms of transfer functions of liquid slosh response, and base moment exerted by the liquid on the tank as a result of horizontal base translation and pitch about a horizontal axis. These results are of

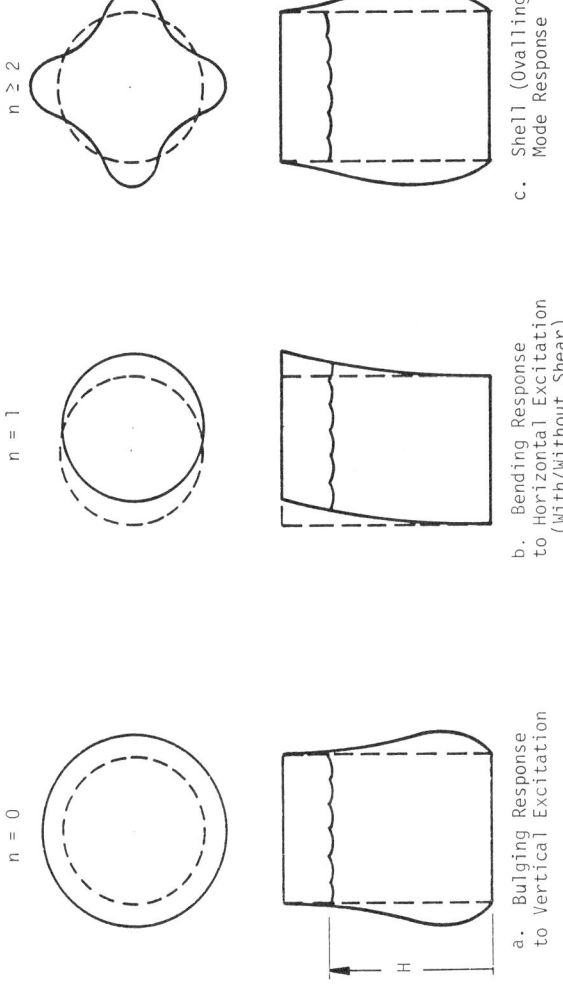

Figure 3. Modal Response of a Flexible Upright Circular Cylindrical Liquid Tank

immediate use when applied with the response spectrum method, i.e., Eqn. (2.2-6), or the power spectral density method, Eqn. (2.2-7). They are especially useful for a quick determination of slosh wave heights for many different tanks. The use of base moment transfer functions in Eqn. (2.2-5) allow an immediate further calculation of maximum slosh longitudinal stress, either by the use of beam or shell equations for the tank. Results are also given for the rigid or impulsive mass for many tanks. This information can be used to predict peak base longitudinal stress due to impulsive loads. However, consideration of tank bending must be included, as will be discussed in Section 3.3.3. In all cases, the peak base longitudinal stress is an important design parameter, since this stress governs the base buckling conditions.

3.3.2 Finite Element Formulations

Finite element formulations are particularly suited for solution of tanks with rather complex geometry. However, this is a rather elaborate approach, and should be used only for those cases where simpler methods are inadequate. Aslam, et al [17] have applied a finite element spacewise solution to sloshing in a rigid toroidal tank and the predicted results compared favorably with those from scale model tests. Generally, impulsive loads can be calculated more readily in complex geometries by the use of the finite element approach, than by the use of the continuum approach. However, slosh responses can often be calculated by the latter method much more simply.

3.3.3 Simple Analytical Models

By far, the most popular method used for design in the past has been the simple equivalent analytical model of the type shown in Figures 4a,b,c. The equivalent mass W_i, spring K_i and damping C_i elements for each slosh mode, and rigid mass W_o are derived from equivalent forces and moments of a liquid exerted on a rigid tank. Expressions for all parameters for models of this type are given in References [3] and [4] for

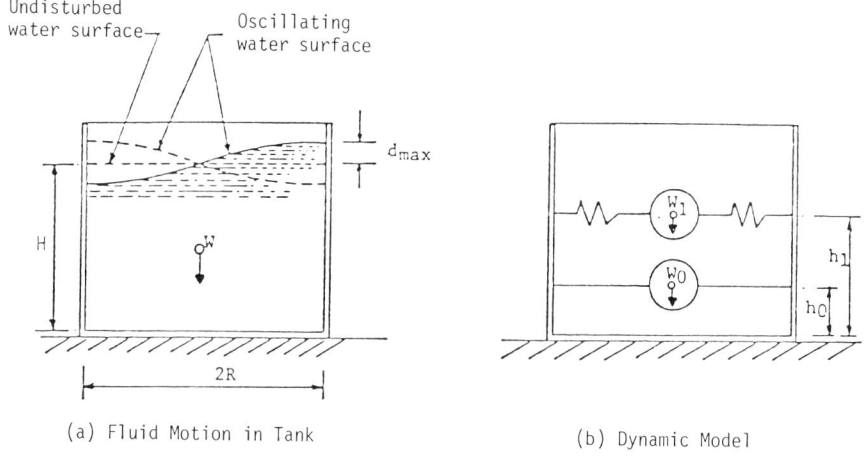

(a) Fluid Motion in Tank

(b) Dynamic Model

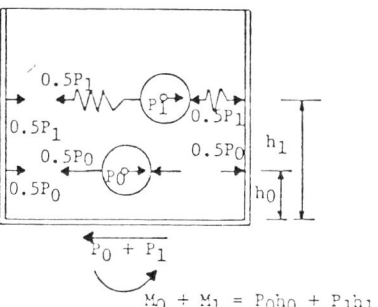

(c) Dynamic Equilibrium of Horizontal Forces

Figure 4. Dynamic Model of Tank with Fluid, Rigidly Supported on the Ground (From Reference 2)

a variety of geometries. They are also given by Housner [1,2] for one slosh mode and rigid mass only for cylindrical and rectangular tanks. Generally, the models are used to predict slosh wave height response, slosh base longitudinal stress, and impulsive longitudinal stress. This is done by assuming that the resulting forces and moments which are induced by a given base earthquake excitation can be applied directly to a rigid beam tank.

The above approach is no longer considered sufficiently accurate for determining impulsive stresses, since elastic container modes are now recognized to occur within the earthquake frequency range. However, a modified Housner approach can be developed, as indicated by the model shown in Figure 5. One approach involves the use of slosh and impulse mass parameters determined in a rigid tank, and placing them in a flexible beam or shell tank. Kalnins [18] has presented the details of this approach. However, Veletsos and Yang [19,50], Fischer [20], Fujita and Shiraki [30], and Balendra, et al [27], have developed an even more accurate approach which accounts for a correction to the rigid mass W_0 which results from the bending deformation of the shell. The latter results are all available only for the upright cylinder configuration. In both of these simple models, the hoop stress can be determined by the effects of the total liquid mass acted upon by the vertical component of the acceleration. However, this approach must also consider the proximity of the breathing mode frequency to the input energy frequency range for maximum accuracy.

3.4 Upright Cylindrical Container

The simple analytical method developed by Housner and others has been the most popular method of design for upright cylindrical containers. Herein a modified Housner approach is outlined, which includes determination of slosh and impulsive masses for a rigid tank, but having them act in a flexible one. The resulting modified model provides more accurate prediction of impulsive responses than the original approach, and is considered essential in modern container design. The method outlined herein is considered at least as a minimum approach to the problem.

3.4.1 Modified Housner Method

This method considers the hydrodynamic effects to be represented as the sum of the convective and impulsive parts, as described in Section 3.2. However, the idealization of the system is as shown in Figure 4 and Figure 5, where elastic tank deformation is allowed. Dimensions h_o and h_1, and the corresponding weights W_o and W_1 are calculated according to the relationships listed in Table 2. The acceleration induces oscillations in which the sloshing portion of the fluid, W_1, responds as if it were a solid oscillating weight flexibly connected to the walls, and the rest of the fluid W_o acts as if it were a solid which is directly attached to the flexible walls. The dimensions h_o and h_1, which locate the resultant forces, also determine the moment at the tank bottom. The overturning moment is computed by increasing these vertical dimensions to allow for the moment of the dynamic fluid forces acting on the bottom. Consequently h_o and h_1 each have two distinct numerical values; the smaller value being used to evaluate the bending moment on a plane just above the bottom, and the larger value being used for determining the overturning moment on a plane just below the bottom. The weight of the tank and the effective weight of the supporting structure are usually small compared to the weight of the fluid. They can be conveniently accounted for by increasing W_o to obtain a modified weight W_o'', which is the gross weight to produce the impulsive force P_o'' (modified P_o value) and the moment M_o'' (modified M_o value corresponding to equivalent weight W_o'').

3.4.1.1 Sloshing Response in Shallow Tanks (H/R<1.5)

Using the appropriate equations from Table 2, the sloshing response can be computed as follows when the tank is supported directly on the ground:

1. Calculate W_1 and the two h_1 values, one excluding the dynamic fluid pressure on the bottom and the other including the dynamic fluid pressure on the bottom.

2. Obtain the natural frequency, ω, from which the quantity S_v, spectral velocity response, may be obtained for the appropriate damping from the design response spectrum curve

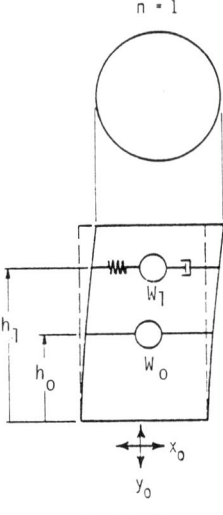

Bending Response

Figure 5. Simple Analytical Model for Base Fixed Flexible Cylindrical Tank

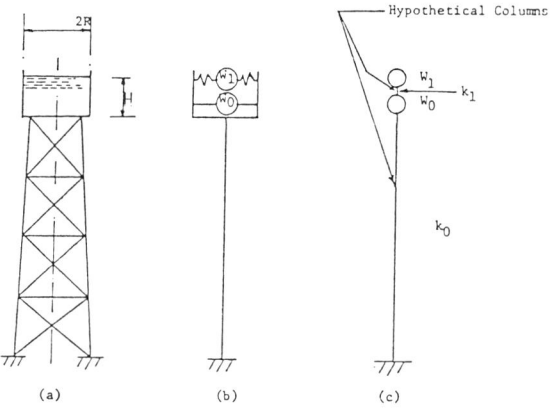

Figure 6. Dynamic Models of Elevated Tanks

TABLE 2. SIMPLE MODEL EQUATIONS FOR CYLINDRICAL TANK
(From Reference 2)

$$C_3 = \frac{\tanh(\sqrt{3R/H})}{\sqrt{3R/H}}$$

$$\frac{W_o}{W} = C_3 \tag{1}$$

$$h_o = 0.375H \qquad \text{(EBP)} \tag{2}$$

$$\frac{h_o}{H} = 0.125(4/C_3 - 1) \qquad \text{(IBP)} \tag{3}$$

$$P_o = \ddot{x}* \frac{W}{g} C_3 = \ddot{x}* w_o/g \tag{4}$$

$$C_4 = 1.84 H/R$$

$$\frac{W_1}{W} = C_5 R/h \tanh(C_4)^{**} \tag{5}$$

$$\frac{h_1}{H} = 1 - \frac{\cosh(C_4) - 1}{C_4 \sinh(C_4)} \qquad \text{(EBP)} \tag{6}$$

$$\frac{h_1}{H} = 1 - \frac{\cosh(C_4) - 2.01}{C_4 \sinh(C_4)} \qquad \text{(IBP)} \tag{7}$$

$$\omega^2 = (1.84 g/R) \tanh(C_4) \tag{8}$$

$$\theta_n = 1.534 (A_1/R) \tanh(C_4) \tag{9}$$

$$P_1 = 1.2 W_1 \theta_h \sin \omega t \tag{10}$$

$$d_{max} = \frac{0.408R \coth(C_4)}{\frac{g}{\omega^2 \theta_h R} - 1} \tag{11}$$

** C_5 = 0.318 according to Ref. (2)
     = 0.354 according to Ref. (34)
     = 0.455 according to REf. (21)

NOMENCLATURE

$W, W_o, W_1, H, h_o, P_o, P_1, d_{max}, R$ - Refer to Figure 4

$\ddot{x}*$ = maximum horizontal acceleration of the ground
ω = circular frequency of free vibration
θ_h = angular amplitude of free oscillations at the fluid surface
A_1 = maximum displacement of W_1, $A_1 = S_v/\omega$
d_{max} = maximum fluid-surface displacement
EBP = excluding bottom pressure i.e., excluding the effect of the dynamic fluid pressure on the tank bottom
IBP = including bottom pressure
g = acceleration due to gravity

for horizontal motion at the level where the tank is located. For liquid slosh in bare wall tanks for damping use $\beta = 0.005$. For any baffle arrangements included, consult Reference 3 for damping values.

3. Using S_v, compute the maximum amplitude A_1 of the W_1 displacement; the angle of free oscillation θ_h at the fluid surface; and the convective force, P_1.

4. Compute the maximum surface displacement, d_{max} (above the original level) from the values of ω and θ_h. Then calculate maximum base shear and moment stress due to slosh response.

3.4.1.2 Elastic Tank Horizontal Mode

Up to this point, the description essentially applies to the original Housner method for development in a rigid container. However, in computing impulsive loads, the container is now assumed to respond in one or more elastic modes, and the impulsive accelerations of W_o are therefore based on spectral acceleration values, rather than the direct ZPA of the base motion. Therefore, an estimate must be performed for the natural frequency of the lateral mode of the elastic tank with rigid mass W_o attached.

Various approximations for estimating the lowest natural frequency for lateral response of the elastic tank and liquid are given in the cited literature [19,21]. Here we present one such approximation which can easily be interpreted in terms of the simple mechanical model shown in Figure 4. The total effective lowest frequency is obtained by a Dunkerley approximation [71] which includes the following individual frequencies:

a) The fundamental cantilever beam bending frequency for the uniform empty tank of height ℓ by itself is obtained from

$$\omega_{11} = \left(\frac{1.875}{\ell}\right)^2 \sqrt{\frac{EI}{2\pi R t_w \rho_s}} \qquad (3.4\text{-}1)$$

b) The frequency of the rigid weight W_o of the liquid, fixed at the end of a weightless cantilever beam of length h_o, having stiffness due to both bending and shear [72] is obtained from

$$\omega_{22} = \left[\frac{3\ EIg}{W_o h_o^3} + \frac{A_s Gg}{2W_o h_o} \right]^{1/2} \qquad (3.4\text{-}2)$$

With these frequencies the combined natural frequency of the tank with concentrated weight W_o can be obtained from

$$\omega_1^2 = \frac{\omega_{11}^2\ \omega_{22}^2}{\omega_{11}^2 + \omega_{22}^2} \qquad (3.4\text{-}3)$$

where

- ℓ = total height of tank
- R = radius of tank
- t_w = thickness of wall
- E = Young's modulus for tank material
- G = shear modulus for tank material
- I = cross-section moment of inertia for tank ($\pi R^3 t_w$)
- A_s = area of tank section ($2\pi R t_w$)
- ρ_s = mass density for tank material.

It can be seen that the weight W_1 of the sloshing part of the liquid does not influence the elastic tank mode. That is, the frequency given by Eq. (3.4-3) is above the sloshing frequency. Note also that the above equations are applicable to a tank of uniform wall thickness. They may also be applied to a tank of nonuniform wall thickness, providing that an equivalent uniform wall thickness can be approximated.

3.4.1.3 Horizontal Impulsive Response (H/R < 1.5)

For thin-wall tanks with values of H/R less than 1.5, reasonable upperbound estimates of the peak values of the impulsive wall pressure and of the associated base shear and moments may be obtained from the corresponding solutions for a rigid tank, merely by replacing the maximum ground acceleration in the expressions for these quantities by the spectral value of the pseudo-acceleration corresponding to the fundamental natural frequency of the elastic tank horizontal mode [19,21].

1. Calculate W_o and two h_o values, one excluding the fluid pressure on the bottom and the other including the fluid pressure on the bottom. Height $h_o = 3/8\ h$ when the effect of fluid pressure on the bottom is ignored.

2. From the above value of W_o and the tank weight, obtain the gross equivalent weight W_o'' and the corresponding h_o'' values.

3. Obtain the impulsive force, P_o'', from W_o'', determining the maximum seismic horizontal spectral acceleration, $S_a(\omega)$, from the design response spectrum at the fundamental structural frequency for lateral motion given by Eq. (3.4-3). A damping value appropriate for the structure should be used.

4. Determine the impulsive bending moment at the base of the tank and the impulsive overturning moment using the h_o'' values obtained in Step 2, above. From this, calculate maximum base shear and moment stress due to impulsive response.

3.4.1.4 Horizontal Sloshing and Impulsive Response in Tall Tanks (H/R > 1.5)

In a circular tank where the depth exceeds three-fourths of the diameter, 2R, the entire mass of fluid below this depth tends to respond as a rigid body as far as impulsive pressures are concerned. For purposes of evaluating the impulsive force, P_o, the container can be regarded as a tank with fictitious bottom at a datum 1.5R below the fluid surface and supported on a solid mass extending from the fictitious bottom to the actual bottom. The procedure of Section 3.4.1.3 (Steps 1 and 2) is applied to the portion above the datum to determine unsprung weight W_o and levels h_o above the datum. The action of the lower portion is represented by the actual weight below this datum and is located at the center of gravity. This actual weight and W_o are combined into a single weight W_o'' at an arm h_o'' above the actual bottom of the tank, with W_o'' generating the impulsive force P_o''.

The concept of dividing the tank into upper and lower zones does not apply in the case of equivalent weight W_1 and related quantities. These continue to be a function of the full depth of water. For determination of the elastic tank effects on the impulsive response, the single weight W_o'' located at h_o'' is used in Eq. (3.4-2) for determining the fundamental elastic tank horizontal mode. However, the effects of the second and higher modes may need to be considered [50].

3.4.2 Vertical Impulsive Response

The modified Housner model defined above does not include any effects due to vertical excitation. Nevertheless, hoop stress at the base of the cylinder should be determined. For a shallow tank this may be obtained from the usual equation

$$\sigma_\theta = p_\ell R \qquad (3.4-4)$$

where

$$p_\ell = \rho g H \ddot{Z} \qquad (3.4-5)$$

However, for a slender tank, this may be insufficient, as pressure amplification can occur because of proximity of the breathing (n = 0) mode to the excitation frequencies. For this case [22]

$$p_\ell = \rho g H S_a(\omega_p) \qquad (3.4-6)$$

where $S_a(\omega_p)$ is the value of the acceleration response spectrum for the excitation at frequency

$$\omega_p = \frac{\pi C_e}{2H} \qquad (3.4\text{-}7)$$

and

$$C_e = \left[\rho \left(\frac{2R}{tE} + \lambda \right) \right]^{-1/2} \qquad (3.4\text{-}8)$$

with t = tank wall thickness
E = elastic modulus of tank wall
λ = liquid bulk modulus.
A damping value appropriate for the structure should be used.

In effect, Eq. (3.4-6) includes the assumption of a closed/open tube standing water-hammer natural mode (n = 0) at frequency ω_p in the liquid/flexible cylinder system. Furthermore, the effect of liquid compressibility may become important in vertical motion response, and is therefore also included. Note also that the vertical impulsive stress calculated by Eq. (3.4-4) should be combined with that caused by horizontal motion by SRSS or another suitable method.

3.4.3 Alternate Simple Analytical Models

The design equations of Table 2 originally derived by Housner have been widely used for predictions of slosh for the first antisymmetric slosh mode and inertial liquid loads. On the other hand, design equations (some of which are somewhat different from those of Housner) can also be obtained from the analytical models given in References [3,4]. The latter equations are more general in that response of any slosh mode (rather than just the first) can be predicted, if necessary. Likewise, expressions for impulsive masses and their vertical locations are also available for use in an alternate form of simple model for an elastic tank. Furthermore, the transfer function method described in Section 3.3.1 can readily be used to predict various responses.

As an example, for application to liquid slosh, the transfer function value for lateral translational excitation at resonance for the rth slosh mode can be obtained as:

$$\eta_r/X_0 = \frac{1}{\beta_r} \frac{\xi_r}{(\xi_r^2 - 1)} \tanh (\xi_r H/R) \qquad (3.4-9)$$

where ξ_r is the eigenvalue for the r-th mode. Substitution of this into Eqn. (2.2-6) results in

$$\eta_r^* = \frac{2\xi_r}{(\xi_r^2 - 1)} \tanh (\xi_r H/R) \, S_d(\omega_r). \qquad (3.4-10)$$

This equation corresponds to that developed by Kana and Dodge [23]. It has been compared with experimental data with Housner's [2] prediction equation (i.e., Eqn. (11) of Table 2) for first mode slosh in a scale model tank by Kana [24]. This comparison is shown in Figure 7; the Housner model tends to become overly conservative at large amplitudes. Expressions similar to Eqn. (3.4-10) for total liquid force and moment about the base can also be readily obtained from the transfer function data given in Reference [3], for both translation and pitching of the tank.

3.4.4 Procedure for Elevated Tank

In the case of the ground-supported tanks previously considered, the weight W_0 or W_0'' was assumed to be elastically coupled to the ground through the tank walls, which experience a peak horizontal acceleration which is amplified over that of the ground itself. When the tank is mounted on an elevated supporting structure, the coupling to the ground will be further complicated, in that the flexibility of the supporting structure also must be considered. The resulting model can be approximated by a two or more degree of freedom system (Figure 6), depending on whether a straight Housner, or modified Housner method is used to represent the liquid and tank. The fictitious springs joining W_1 to the tank walls in Figure 4 have been replaced with a single hypothetical column of the same stiffness, k_1, forming a direct coupling between W_1 and W_0. The weight W_0 is connected to the ground through a similar hypothetical column

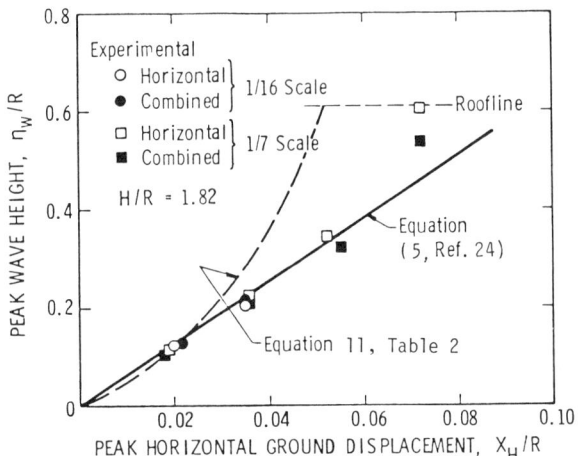

Figure 7. Peak Liquid Responses in Model Storage Tank of Upright Cylinder
(From Reference 24)

Reprinted with permission of Argonne National Lab.

representing the support structure and having the same spring constant, k_o. The properties W_o, W_1, h_o, h_1 and ω are independent of the support conditions and are obtained from Table 2. Frequencies, mode shapes, maximum deflections, etc., of the system can be found for each mode. The contributions from both modes are summed so as to maximize values of force and deflection. In computing the maximum vertical displacement, d_{max}, of the fluid surface contributed by the second mode, the following expression is used rather than Equation 11 in Table 2.

$$d_{max} = \theta_h R$$

Damping values associated with sloshing are approximately 0.5% of critical, whereas damping associated with the motion of the support may be considerably larger. An example of the use of the Housner method applied to an elevated tank has been given by Lee and Reddy [49].

3.4.5 Consideration of Base Fixity and Buckling

Calculation of peak base longitudinal stress due to both sloshing and impulsive bending must be performed to determine the potential buckling failure of the tank. However, all of the above procedures apply to a fixed base tank. Much less is known about base-free tanks where uplift can occur. Experiments by Clough [25] show that the buckling problem appears to become more acute for the base-free case. More recent observations of Niwa and Clough [51] indicate that shallow ($H/R < 2.0$) base-free tanks tend to buckle at the base in an "elephant-foot" form, while slender tanks ($H/R \approx 4.0$) tend to buckle in a diamond-shaped pattern spread around the circumference. Examples of this behavior are shown in Figures 8A and 8B respectively, where such buckling occurred in tanks of various sizes at the Wente Bros. winery during the Mt. Diablo earthquake of January 24, 1980. After performing a series of experiments on similar tanks, Niwa and Clough suspect that elephant-foot buckling results from the combined action of vertical compression exceeding the critical stress and hoop tension occurring near the yield limit.

Figure 8a. Elephant-foot Buckle in Wente 'broad' Tank
(Reference 51)

Figure 8b. Diamond-shaped Buckle in Wente 'tall' Tank
(Reference 51)

Reprinted with permission of Earthquake Engineering and
Structural Dynamics

3.5 Sloshing in Tanks of Various Geometries

For tanks of various other geometries, there is in general less design information available in the open literature. Therefore, in this section we will identify sources of information that are available, and provide some discussion of results where appropriate. In most cases, only data for slosh response in rigid containers are available.

3.5.1 Rectangular Tank

Slosh response in rectangular tanks can also be predicted by means of simple analytical models. Generally, the equations of Housner [2] have been used for designs in which only the fundamental slosh mode is included. However, similar simple model equations are also available in References [3,4]. Furthermore, using the transfer function approach, additional modes can readily be included where necessary. As shown in Figure 9, for a rectangular tank of width "a" in the direction of horizontal excitation, the transfer function for maximum slosh in mode "r" can be determined [3,4] as

$$\eta_r/X_o = \frac{2}{\pi\beta_r(2r+1)} \tanh\ [(2r+1)\pi H/a]. \tag{3.5-1}$$

Note in this case $r = 0$ is the first mode. Substitution of this into Eq. (2.2-6) results in the maximum wave response in mode r

$$\eta_r^\star = \frac{4}{\pi(2r+1)} \tanh\ [(2r+1)\pi H/a]\ S_d(\omega_r) \tag{3.5-2}$$

If higher mode responses become important, the results can be predicted and combined by SRSS. Similar expressions can be determined for the tank wall loadings using the transfer function technique. No published information is available for design of liquid slosh in flexible rectangular tanks. However, it would appear that an approach similar to that described in Section 3.4 for the cylindrical tank would be adequate.

3.5.2 Horizontal Cylindrical Tank

This confiruration is used for liquid storage purposes at some nuclear plant installations, and is shown in Figure 10. Some experimental

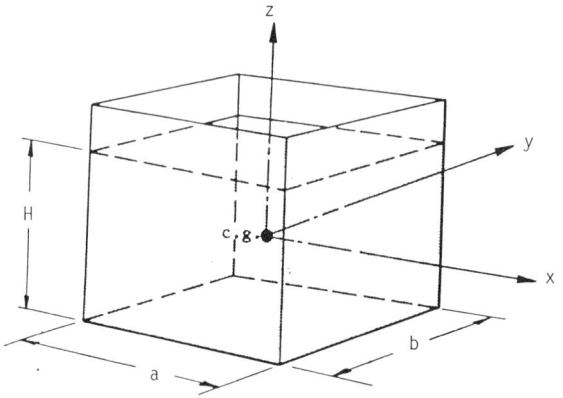

Figure 9. Coordinate System for Rectangular Tank

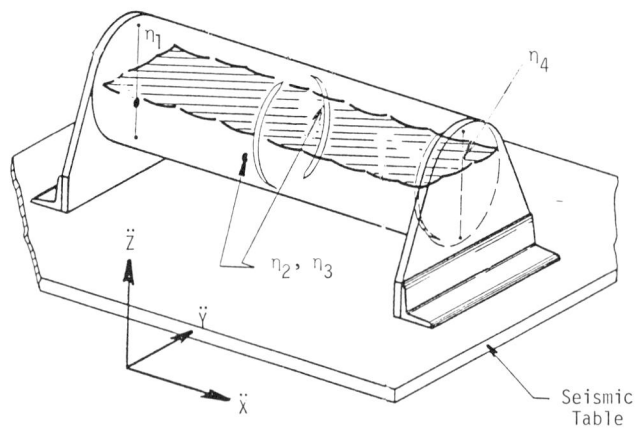

Figure 10. Excitation Arrangement for Horizontal Cylindrical Tank

work to determine natural frequencies has been reported in one aerospace monograph [3]. Apparently, analytical transfer functions for this configuration have not previously been determined. The geometry is complicated by the cross-sectional curvature, as well as the possibility of significant length differences longitudinally compared to laterally. Thus, essentially two different slosh configurations exist along the two principal tank axes, so that for seismic response, the effects of two independent horizontal excitations must be accounted for.

A study of seismic slosh in this configuration was reported by Kana [26]. Experimental data were obtained for a series of simulated earthquake runs under various liquid depth and tank orientations relative to the horizontal excitation. The results were compared to predictions made by equations based on a rectangular tank of equal dimension at the liquid level and in the direction of excitation, and of a depth which allowed equal liquid volumes. Figures 11a and 11b show some comparison of experimental results with these predictions. The latter results are based on the equations of Housner [2], and also on Eq. (3.5-2) of the previous section, in which only the first mode is included. It can be seen that the two methods more or less bracket the data from the transverse axis and agree for the longitudinal axis.

The results predicted by Eq. (3.5-2) do not appear in Reference [26]. The ease by which such predictions can be made from previously available analytical transfer functions was realized only after that work was completed. However, predictions were made by a semi-empirical method, whereby experimentally measured modal damping and transfer functions at resonance were used in Eq. (2.2-6) to predict results under a variety of conditions. Examples of these results also appear in Figure 11. However, even this approach was complicated for some depths in the horizontal cylinder because of the presence of strong nonlinearities. At best, it can be seen that a purely analytical approach to prediction of response for this geometry will require taking account of the nonlinearities present if moderate to large excitation is to be considered. Such ranges of excitation are not uncommon as requirements of qualification of some nuclear plant facilities.

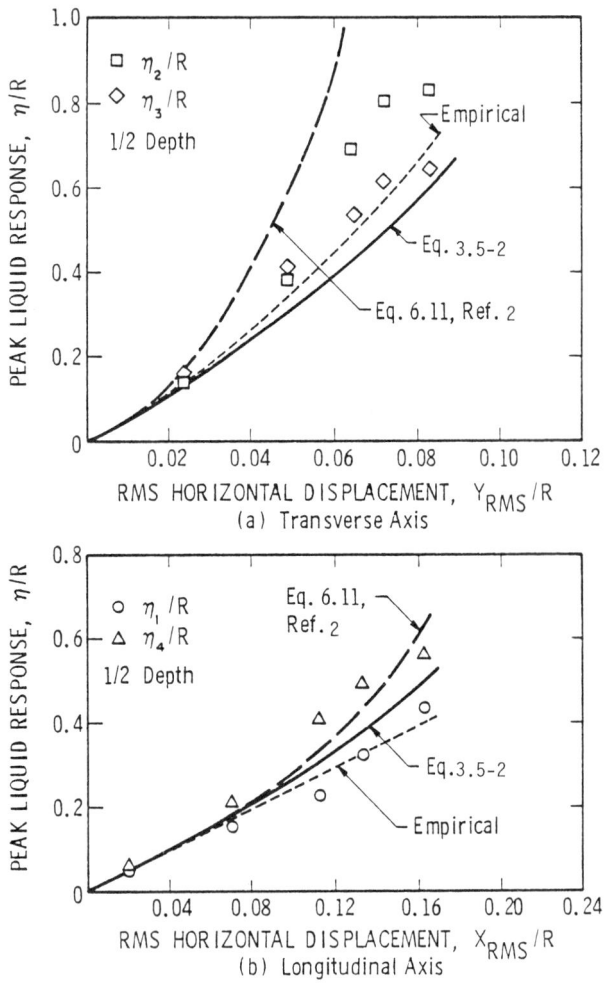

Figure 11. Liquid Slosh Response for Simulated Seismic Excitation of Principal Axes in Horizontal Cylinder (From Reference 26)

3.5.3 Annular Cylindrical Tank

The annular tank geometry has been used for the steam suppression pool of some BWR nuclear power plants, and as such, prediction of seismic slosh response is very important to assure proper operation of the system.

Aslam, et al [28,29] have reported a combined analytical and experimental effort which resulted in a digital computer time history method of predicting the response to specified earthquake accelograms, and comparing the results with a series of measurements in a scale model tank. The experimental results provide valuable data for comparison with analytical models. Nevertheless, the same transfer function and response spectrum approach previously described can readily be applied to this configuration from already available data as will now be shown from Figure 12.

From either of the two monographs [3,4] the transfer function at r^{th} resonance for wave response at the inner wall at $r = b$ in line with lateral translation of a rigid annular tank of outer radius $r = a$ can be found as

$$\eta_r/X_0 = \xi_{r1} \frac{\overline{A}_r C_1(\xi_{r1}b/a)}{2\beta_r} \tanh(\xi_{r1}H/a) \quad (3.5\text{-}3)$$

where ξ_{r1} are roots of a frequency determinant and are tabulated in Reference [3], and $\overline{A}_r$ and $C_1(\xi_{r1}b/a)$ are rather complicated expressions that can be evaluated for a given tank by the information given in Reference [3]. In this case, $r = 0$ is the first slosh mode. By substitution of this expression into Eq. (2.2-6), the peak response at the inner wall caused by the rth mode can be obtained as

$$\eta_r^* = \xi_{r1}\overline{A}_r C_1(\xi_{r1}b/a) \tanh(\xi_{r1}H/a) S_d(\omega_r). \quad (3.5\text{-}4)$$

Results from this equation were calculated for the second slosh mode in the same annular tank investigated by Aslam, et al [29], and compared with their data as shown in Figure 13. The experimental measurements were taken from Table 5-2 of Reference [29]. The predicted time

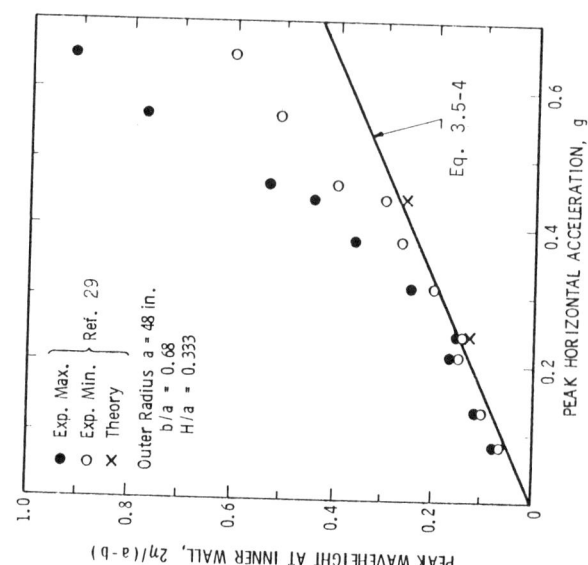

Figure 13. Seismic Slosh Response in Annular Tank Under El Centro, 1940 Earthquake (From Reference 5)

Reprinted with permission of Argonne National Lab.

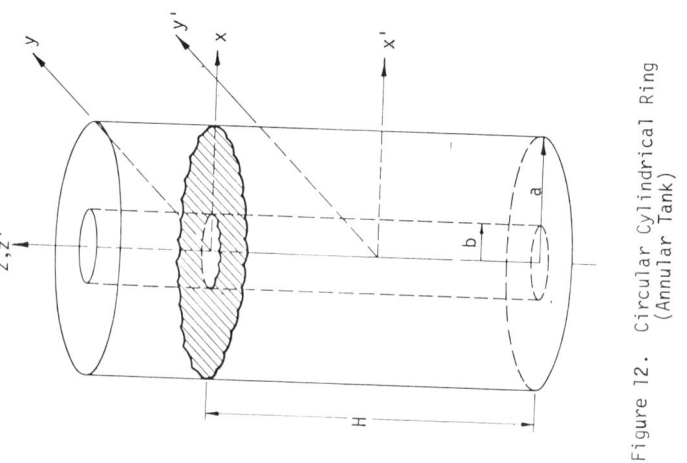

Figure 12. Circular Cylindrical Ring (Annular Tank)

history results were read from Figures 5-32 and 5-34 given in the report. Calculations based on Eq. (3.5-4) included known values of the response spectrum for the N-S component of the May 18, 1940, El Centro earthquake whose excitation was applied in the study. It can be seen that the use of Eq. (3.5-4) provides results comparable to the much more complex time history approach. Furthermore, complete expressions for force and moment loadings for a simple analytical model are available for the annular tank by using the transfer function and response spectrum approach. It is also apparent from Figure 13 that neither theoretical approach is adequate beyond about 0.3 g excitation.

One other aspect of the annular tank results should be emphasized. In this case, the second mode turns out to provide the maximum seismic response and it occurs at the inner rather than the outer wall. This shows the danger of always assuming that the first mode provides maximum response and the utility of having a method available whereby response in any mode can be determined.

3.5.4 Toroidal Tank

The toroidal configuration is used for steam suppression chambers in many operating BWR nuclear power plants. For this application, typically the toroidal tank contains water to about 1/2 depth, similar to the diagram in Figure 14. Determination of liquid slosh is very important to the operation of these systems. A recent study by Aslam, et al [17] has included an equivalent annular tank development as described in Figure 13 and a finite element development both combined with a time history solution for response prediction of waveheight and pressures for this configuration. Some results from this study for a 1/60-scale tank are shown in Figure 15. Here the peak waveheight response is plotted against displacement, rather than acceleration since, as pointed out in the reference, the frequency content of the time history was distorted from a true El Centro reproduction because of filtering action in the shaker table. It can be seen that the time history solution provides an adequate prediction up to about $X/R = 0.03$. Beyond this, nonlinearity of the response dominates.

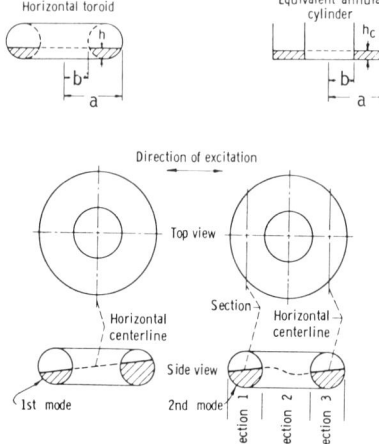

Figure 14. Calculated and Experimental Values of the First Two Natural Frequencies for Horizontal Orientation of Toroid (From Reference 3)

Reprinted with permission of Southwest Research Institute

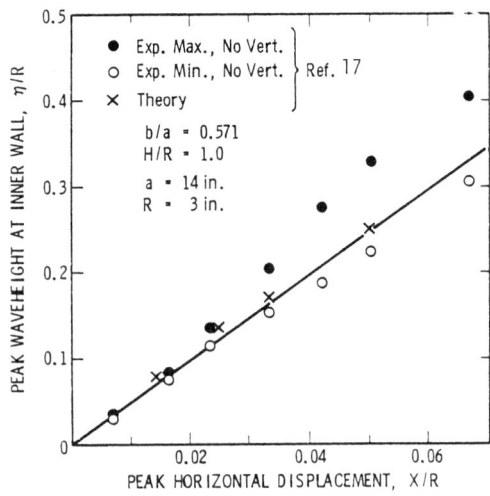

Figure 15. Seismic Slosh Response in Torus Tank Under Modified El Centro, 1940 Earthquake (From Reference 5)
Reprinted with permission of Argonne National Lab.

As was the case with the annular tank, it may also be suspected that for the torus, the rather complex time history method is unwarranted. An equivalent volume response spectrum solution can also be applied for the 1/60-scale tank by using Eq. (2.2-6) and comparing the predictions to the results shown in Figure 15. However, this could not be done from the data presented in Reference [17], because of the distortion of the time history from the true El Centro earthquake. Eq. (2.2-6) could be applied directly if the response spectra from the experimental time histories were available.

The spectral approach should produce waveheight predictions with the same accuracy as the time history solution. However, as pointed out by Aslam, et al [17], the equivalent tank method probably does not provide good predictions of impulsive pressures because of the inherent difference in geometries. The finite element approach is necessary in this case. Note, however, that even with a finite element solution to the spatial part of the problem, the spectral method can still be employed for the time dependent response.

3.5.5 Other Geometries

There are still several different geometries for tanks that can be used for industrial facilities. Some data for liquid slosh response in most of these configurations is also included in the aerospace monograph [3]. Seismic slosh in a spherical tank has been studied by Nakano and Watabe [31]. This, of course, is a special case of an ellipsoidal tank. Furthermore, there is a variety of data available for liquid slosh in cylindrical sectored tanks of various configurations including design equations for forces and moments and representation by simple analytical models. In many cases, equations are also available for pitch excitation as well as lateral translation at the base.

4.0 SUBMERGED STRUCTURES - ADDED MASS AND DAMPING

4.1 Overview

A number of structures, such as spent-fuel storage racks, main pressure-relief valve lines and internals of the reactor vessel are submerged in water. For these structures, the effect of the water in terms of forces and damping must be considered. These effects for the case of submergence in a liquid are accounted for by increasing the mass and damping of the structural systems. A survey of the state-of-the-art in this area has been presented by Dong [6]. Two theories are used to account for the dynamic effect on submerged structures - (i) incompressible inviscid theory [32] (potential theory), and (ii) compressible inviscid theory [33]. The incompressible inviscid theory is used for non-flexible members (i.e., members that can be treated as translating rigid bodies). The compressible inviscid theory is used for flexible members, such as flexible coaxial cylinders. There is good agreement between potential theory and experimental data which leads to the conclusion that potential theory satisfactorily describes the added mass phenomenon.

There are two types of submerged structures that need to be considered under this section. They are single isolated members like main pressure relief valve lines and multiple members like spent-fuel storage racks and internals of the reactor vessel which are submerged in water. The dynamic effects of water for these two types of members will be discussed below.

4.2 Single Isolated Members

In most cases of practical interest, the earthquake-induced vibrations of submerged structures can be studied satisfactorily under the assumptions that the velocity of the structure relative to the surrounding fluid is sufficiently low that the liquid may be taken as incompressible, inviscid, and irrotational. Under these conditions, the phenomenon can be analyzed easily by adding, to the mass of the structure (not considering the buoyant effect of the liquid), the mass of a certain volume of liquid, which gives a total "virtual" mass. Because the submerged structures tend to translate with the pool, the surrounding water will decrease the relative motion. This anchoring effect can be represented by modifying the external forcing matrix in Eq. (2.1-26) to reflect the buoyant weight in air. This will lead to realistic responses, especially for pools with rigid walls.

If the structure is a long, slender, rigid prism on flexible supports, moving in a direction perpendicular to its axis, flow of liquid around the structure is essentially two dimensional. Under these conditions, the added mass is that of a circular cylinder of liquid having the same length as the prism and a diameter equal to the width of the projection of the prism on a plane perpendicular to the direction of motion as shown [34] in Figure 16. Experiments by Clough [35] have shown that the additional mass may be on the order of 25 percent greater than this value for rigid prisms on flexible supports, and slightly less than the theoretical value for long, flexible prismatic structures. In addition, damping due to liquid viscosity may be disregarded because, even if energy dissipation due to radiation into the liquid may be more important, the model tests of Clough [35] indicate that it will not exceed about 2 percent of critical for submerged structures of ordinary dimensions.

With this brief summary, details of computation of added mass and damping will be described below.

4.2.1 Added Mass for Single Isolated Members

The accelerating fluid in a fluid-structure system induces an added mass effect onto the submerged member. Under sufficiently small amplitudes of motion, whether cyclic or unidirectional, the added mass phenomenon can be described in terms of an added mass coefficient C_m defined as

$$C_m = \frac{\text{added mass of fluid}}{\text{reference fluid mass}} \qquad (4.2-1)$$

where the reference fluid mass is that of the cylinder of fluid of diameter

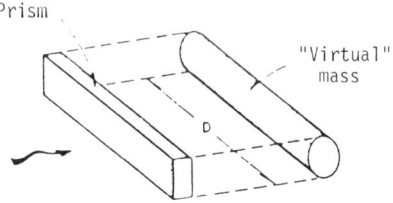

Figure 16. Submerged Body and Its Virtual Mass

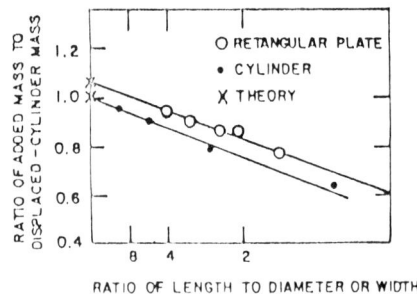

Figure 17. Circular Cylinder and Rectangular Plates
(From References 6, 38)

equal to the dimension perpendicular to the direction of motion, or in some cases, it is the mass of the displaced fluid. The added mass phenomenon for single isolated members has been rather extensively investigated experimentally and analytically [36, 35, 37, 38]. Theoretical treatment based on potential theory has been found to be adequate.

The value of C_m for a fluid moving around a stationary specimen is higher than the value for a specimen moving in a stationary fluid as exhibited theoretically [39, 40], as well as experimentally and is important to account for in real applications. For a stationary circular cylinder in a moving fluid, such as a single fuel bundle element in a vessel, $C_m = 2$, which means the hydrodynamic force acting on the stationary cylinder is twice the mass of fluid displaced times the acceleration of the fluid. By comparison, for a translating circular cylinder in a stationary fluid, such as submerged piping in a suppression pool, $C_m = 1$, which means the total force required to accelerate the cylinder is the mass of the cylinder plus the mass of the displaced fluid multiplied by the acceleration of the cylinder. For the case in which both the cylinder and fluid are in motion, these two force contributions should be calculated separately and superimposed.

For a member with finite length, the fluid flows around the end(s) as well as around the length. Therefore, the inertial resistance to motion per unit length is less than that for an indefinitely long member. Figures 17 and 18 [Ref. 38] illustrate the effect experimentally and theoretically for specimens with both ends free for fluid to flow around. These curves could be applied in an approximate sense to cross-sections other than those shown in the figures.

When a member is only partially submerged in water, the added mass effect decreases near the water surface [34]. The decrease in total added mass for member of radius r, as a function of depth, h, of submersion is given in terms of a correction factor in Figure 19.

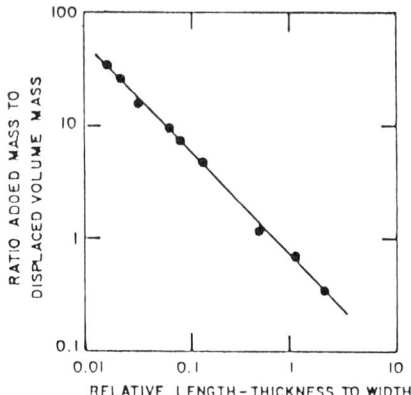

Figure 18. Relative Effect of Virtual Mass in Parallelepipeds
Square Side Moving Broadside On
(From References 6, 38)

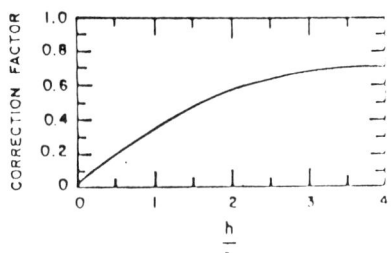

Figure 19. Liquid Mass Correction Factor in
Circular Pier
(From References 6, 34)

4.2.2 Added Damping for Single Isolated Members

The damping force acting on a submerged member is usually relatively small and not included in the analysis as an acting force. Instead, the effect is usually described as an equivalent viscous damping. The contributions to added damping are: 1) Fluid Viscosity, 2) Component Impact, 3) Wave Generation, and 4) Acoustic Generation. The last two are forms of radiation damping (i.e., wave and acoustic energies generated radiate away from the submerged members). A significant amount of acoustic energy generation is not expected for the structures and excitations under consideration. Wave generation is generally not important for fully submerged structures under seismic excitation [35, 34]. For partially or fully submerged structures in a finite-sized water enclosure, radiation damping is usually not taken into consideration because the radiation energy may bounce off the enclosed walls back to the submerged member. Component impact may be a significant source of damping for only multiple members and not for single members. Therefore, for single isolated members fluid viscosity is the only source of damping in need of consideration. Experimental data given by Skop, et al [37] indicate that as the amplitude (specimen displacement) diameter ratio (A/D) increases, the type of damping changes from viscous damping to a nonlinear damping where the damping force becomes proportional to the square of the velocity. Figure 20 illustrates the relationship where A/D values change from linear to nonlinear damping for different specimen diameter. A gradual increase with diameter is seen. Nonlinear damping is seen to be greater than the linear damping, so that the use of the linear damping value as an approximation in the nonlinear range will be conservative. In addition, experimental data [36, 35, 37, 41] also indicate that added damping decreases with increasing specimen size.

In structural analysis, damping is expressed as either a damping coefficient or a percent of critical damping. Damping coefficient describes the damping independent of the mass and stiffness of a structure, whereas the percent of critical damping is a description associated with the mass and stiffness. To see which description best fits the added damping

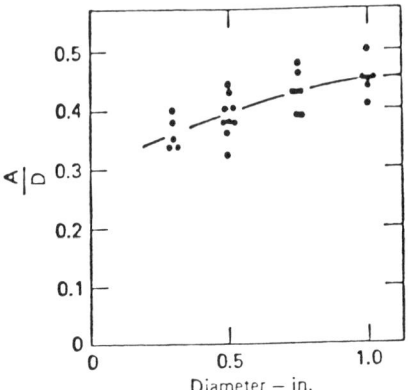

Figure 20. Amplitude/Diameter Value for Linear Damping Oscillating Submerged Circular Cylinders (From Reference 6), Reprinted with permission of Lawrence Livermore Lab.

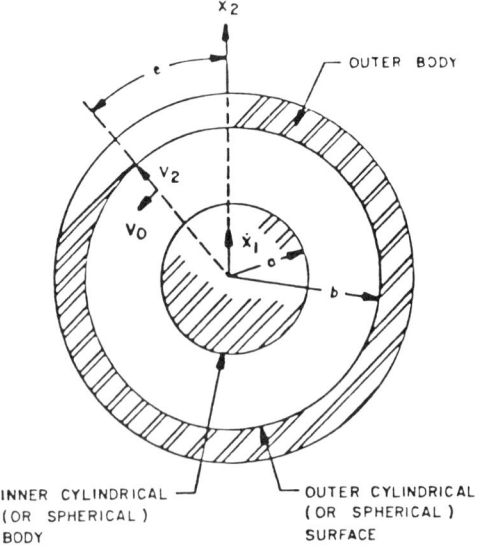

Figure 21. Two-Body Motion with Fluid Coupling

from water, the data given by Skop, et al [37] were converted by Dong [6] to both an added coefficient and an added percent of critical damping. The results showed that added coefficient varied in an inconsistent manner with frequency but the added percent of critical damping varied less and was more consistent. Hence, it is reasonable to treat added damping as an addition to percent of critical damping independent of frequency of vibration under consideration.

Based on these discussions and on the data from References [36,35, and 41], added damping values for some of the typical single isolated components are given in Table 3.

4.3 Multiple Members

When a group of members submerged under water is subjected to earthquake loads, the fluid dynamic effects on these members are more complex than for a single isolated member. This is because of the arrangement of the members, space between them, motion of one member relative to another, and the generation of lift forces. Several highly theoretical investigations are available, [42,43,46,52,53], some of which are rather complicated to be of practical use in design analysis. These investigations are classified into three types of arrays: 1) Groups of cylinders; 2) Groups of cylinders inside a large circular cylinder; and 3) Coaxial flexible cylinders. They will be discussed briefly in the following subsections.

4.3.1 Hydrodynamic Coupling for Groups of Cylinders

Closed form solutions using potential theory are available in References [42,43,40, and 44], where the solution is expressed in terms of "self added" and "added" mass coefficients. The self added mass coefficients characterize the hydrodynamic forces on a member from the motion of the member itself with all other members held stationary. The added mass coefficients characterize the hydrodynamic forces in a stationary member with other members in motion. Experimental comparisons with values

TABLE 3. ADDED DAMPING VALUES PROJECTED IN
FIGURE 16 FOR SINGLE, ISOLATED STRUCTURES
(FROM REFERENCE 6)

Structure	Size, in.	Added Damping % of Critical
Fuel elements	~ 0.5 D	≤ 4.2
BWR fuel bundle	~ 5.5 × 5.5	≤ 0.55
PWR fuel bundle	~ 10 × 10	≤ 0.33
Main steam-relief valve line	8 D	≤ 0.40
	12 D	≤ 0.29

obtained using potential theory are given in References [42 and 40] for a seven-member hexagonal array and a 3 x 3 square array which show that agreement between theory and experiment is good. Although experimental confirmations such as the above are few, they are generally good for arrays of cylinders. Combining this with the excellent confirmation established for single isolated members, it can be concluded that potential theory can be used adequately to describe the added mass and lift (transverse Karman vortex induced) forces for groups of cylinders.

Where there is a large array of cylinders under consideration, the most significant hydrodynamic coupling for a given member is with its immediate neighbors; therefore, coupling with members farther removed can be neglected [43,6]. Hence, regardless of the size of the array, it can be analyzed in sub-parts consisting of each member and its immediate neighbors. However, care must be exercised when evaluating forces on corner members.

4.3.2 Hydrodynamic Coupling for Rigid Members Surrounded by a Rigid Circular Cylinder

The model commonly used to simulate the internals of the reactor vessel (of a BWR) under seismic excitation is that of the rigid coaxial cylinders. For two rigid coaxial cylinders in motion, as shown in Figure 21, with the annular space filled with fluid, the solution using potential theory is expressed in a form quite convenient for design applications. The fluid forces on the inner and outer cylinders are, respectively [33],

$$F_{f_1} = -M_H \ddot{X}_1 + (M_1 + M_H) \ddot{X}_2 \qquad (4.3-1)$$

$$F_{f_2} = (M_1 + M_H) \ddot{X}_1 - (M_1 + M_2 + M_H) \ddot{X}_2 \qquad (4.3-2)$$

where,

$M_1 = \pi a^2 L \rho$ = mass of fluid displaced by the inner cylinder

$M_2 = \pi b^2 L \rho$ = mass of fluid that could fill the outer cylindrical cavity in the absence of the inner cylinder

$M_H = M_1 \dfrac{b^2 + a^2}{b^2 - a^2}$ = mass term depending on the relative sizes of the inner and outer cylinders.

L = length of the cylinders, ρ = mass density of fluid.

Values for M_1 and M_2 can be determined experimentally or theoretically. (Additional correction factors are applied to M_1 and M_2 to account for leakage of fluid from the fuel cell and axial fluid flow through the fuel cell). These equations theoretically apply only to infinitely long cylinders; therefore, L should be significantly larger than the radii a and b. In addition, the solution's validity should diminish if the annular space is very small compared to radii a or b because the fluid would then be subjected to a significant amount of flow and shearing to accommodate the relative motions of the cylinders. The incompressible and inviscid assumptions would then be less valid.

The finite element technique can also be used to evaluate dynamic effects of submerged coaxial cylinders. Results presented by Levy and Wilkinson [45] and Wu and Levy [46] include the use of a finite element method, and compare reasonably well with potential theory in terms of added mass coefficients for the case of two coaxial rigid cylinders. The basis of comparison was the M_1, M_2, and M_H of Eqs. (4.3-1) and (4.3-2). A somewhat more sophisticated treatment of coaxial rigid cylinders is given by Chen, et al [41] and Fritz and Kiss [47], using an incompressible viscous theory. The solution expressions are much more complex than those for potential theory and are contained in those references. A comparison with experiment was made [41] for a fixed outer cylinder and oscillating inner cylinder. The outer cylinder diameter was varied from 0.625 in. to 2.5 in., while the inner cylinder diameter was kept at 0.5 in. The agreement between analysis and experiment was found to be quite good and noticeably better than the comparisons found for the potential theory. A possible conclusion is that viscous effects may be important and perhaps should be included when analyzing coaxial rigid cylinders. More experimental comparisons are needed to confirm this possibility.

4.3.3 Hydrodynamic Coupling for Flexible Coaxial Cylinders

Coaxial cylinders with the inner cylinder analyzed as a flexible shell may probably provide more realistic modeling of the internals of the reactor vessel than would coaxial rigid cylinders. Such a mathematical model has been analyzed using incompressible inviscid theory and the deformation of the inner cylinder compared reasonably well with experimental results given by Au Yung [48]. However, in view of the very limited number of experimental comparisons, the compressible inviscid potential theory appears to do better than the incompressible inviscid potential theory. This indicates that fluid compressibility may be quite important to include when analyzing flexible members. Additional experimental confirmation is needed to fully establish this possibility.

4.3.4 Damping for Multiple Members

In previous discussions, it was explained that for fully submerged structures in a finite size container, radiation damping can generally be ignored. Hence, the only contributions to added damping are from fluid viscosity and component impact. Both theoretical and experimental values for fluid viscous damping have been published, although no analytical treatment of impact damping has been found. For experiments involving both fluid viscosity and component impact, no separation of the measured total damping into these two contributions has been made. Establishing a fixed value of damping for a general multiple member structure is very difficult, if not impossible, because damping can be significantly influenced by member arrangement, spacing, and relative motions among the members. Thus, any arrangement must be evaluated experimentally to determine its damping characteristics.

5.0 FLOATING STRUCTURES

5.1 Overview

Orr and Dotson [64], and Dean, et al [74] have described the concept of a floating nuclear plant (FNP) of the Atlantic Generating Station (AGS) type, as shown in Figures 22 and 23. Although this concept has been shelved for the time being, it is not inconceivable that a similar concept would be revived in the foreseeable future. In this section the basic, conceptual, and computational features are outlined. The topics covered are (i) a brief description of the AGS power plant, (ii) amplification of water-transmitted seismic excitations, (iii) determination of the protective basin response, (iv) fluid-structure interaction analysis of the floating platform, including the effect of basin oscillations, (v) determination of the response of the mooring caisson and the mooring strut.

In view of the considerable interest in floating concepts by the oil industry, namely the Tension Leg Platform (TLP shown in Figure 24), and the Liquified Petroleum Gas (LPG) storage platform, there is a lot of potential for the transposition of this technology to the FNP. Work on Ocean Thermal Energy Conversion (OTEC) platforms will also provide important spin-off benefits to offshore nuclear plant concepts.

5.2 AGS Power Plant

The proposed AGS power plant was a steel barge, 400 ft square and 44 ft deep, reinforced with bulkheads to form a honeycomb of watertight compartments. The domed containment was nearly 18 stories high above the ocean surface. The breakwater was a massive curved structure of 5.6 million tons of stone and cast concrete, spanning 49 acres of ocean floor with a rise of 64 ft above the water surface. The power transmission cable connecting the plant to the power distribution grid was 4 miles long. The plant was to use a pressurized water reactor (PWR) to generate 1150 MW(c) of power with a design life of 40 years.

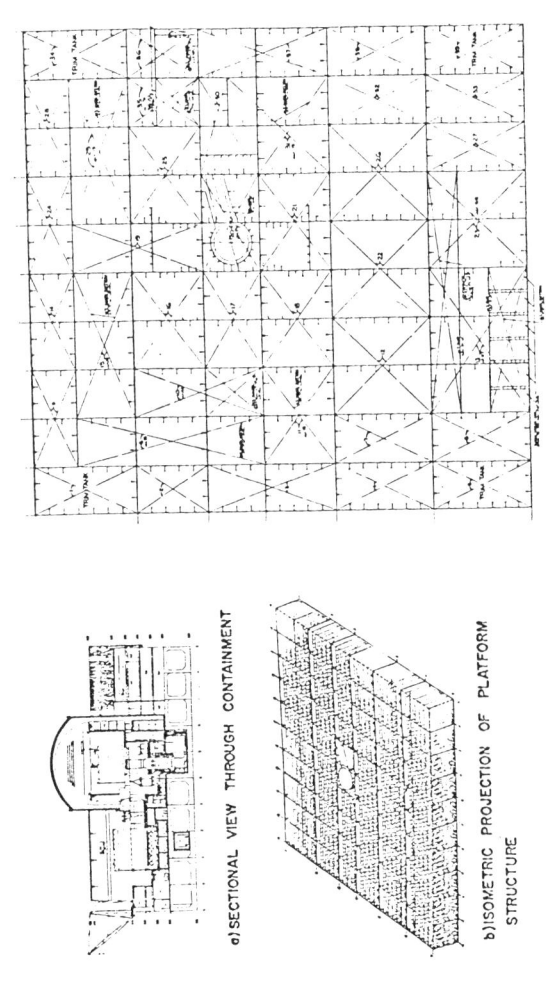

Figure 22. Atlantic Generating Station (Reference 64)

Reprinted with permission of Argonne National Lab.

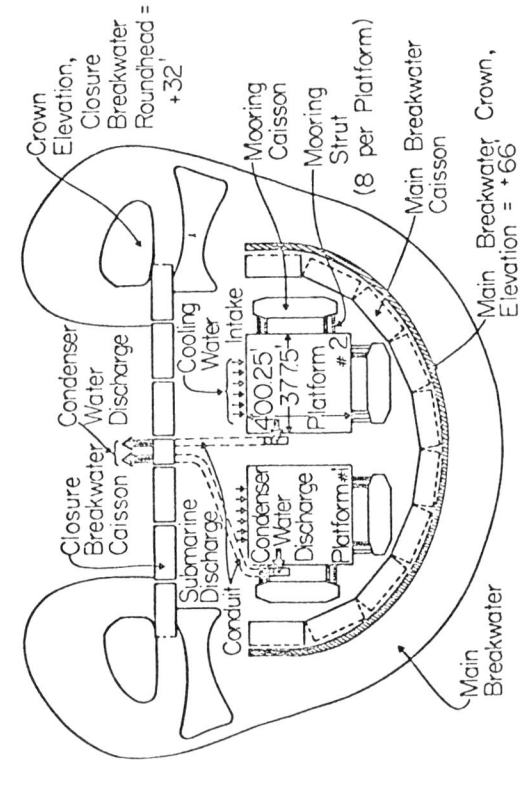

Figure 23. Features of the Atlantic Generating Station (Reference 74)

Reprinted with permission of the Offshore Technology Conference

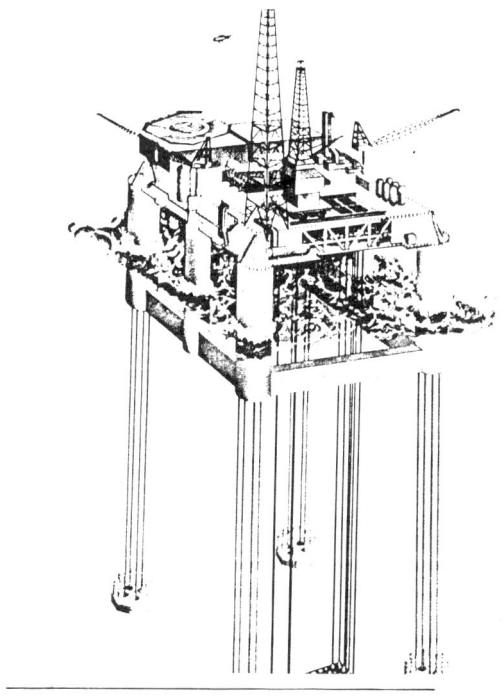

Figure 24. Concept of a Tension Leg Platform

5.3 Amplification of Water Transmitted Seismic Excitations

Contrary to popular thinking that the FNP concept minimizes the seismic influence in view of the shear free water medium, the supporting water column significantly amplifies the vertical component of the earthquake even in shallow water depths. In some of the recent work presented by Thangam Babu and Reddy [56,60] and Thangam Babu [63], the amplification factor for a Taft Earthquake input was about 30 for 40 ft water depth. The problem becomes even more complex for moderately deep water with the phenomenon of cavitation. Large negative over-pressures due to the incident and reflecting (tensile) waves cause cavities to form. This results in a 'slamming' down effect that induces large stresses in the floating concept. The effect was first observed in ships, and preliminary dynamic analysis performed by Williamson et al [65] confirmed this. The idealization was a simple lumped spring (bilinear tension cut-off) dashpot-mass model. The cavities were simulated by a 'gap' opening at constant tensile stress. The closure of the 'gap' was treated as elastic impact for the sake of conservative analysis.

5.4 Basin Response

The natural frequency and mode shapes and the dynamic amplification factors of the basin can be evaluated once the geometry of the basin, platform boundary, and the frequency and heading of the incident wave are known. The boundary value problem for the basin alone can be described by the Laplace equation in terms of the velocity potential ϕ. Using Green's theorem, the potential, $\phi(x,y)$, can be numerically determined following Hwang, Li-Sam, and Tuck [66], after satisfying the kinematic and rigid basin boundary conditions. For a more rigorous analysis, the wave scatter effect should be included by expressing the velocity potential as the sum of the velocity potential of the incident wave, ϕ_o, and the velocity potential of the scattered wave, ϕ_s. This requires that the platform structure be included in the overall basin model. Furthermore, for varying shallow depths, the wave refraction and diffraction effects should also be incorporated. Bettess and Zienkiewicz [67] have taken into account these factors and used finite and infinite elements to evaluate the amplification factors of the surface elevation for the AGS site. The fluid domain was divided into inner and outer domains. Finite element discretization with standard variational formulation was used in the inner region, and infinite elements in the outer domain.

5.5 Fluid-Structure Interaction Analysis

Two basic formulations, namely, the Eulerian and Lagrangian approaches, have been used in solving problems in this category. In the Lagrangian method, the fluid displacements are used in the same manner as the structure, and the fluid is treated as an elastic medium with a small but finite shear modulus. The Eulerian approach uses fluid pressure as the unknown and the coupling is achieved by evaluating the solid/fluid interface forces. This method is widely used due to the decreased number of unknowns involved in describing the problem. The finite element formulation for the general case of fluid-structure interaction problems which takes into account the incompressibility of the fluid, the flexibility effect of the structure, surface wave effects, wave radiation damping effects, and scattered wave effects is presented below.

5.5.1 Structure Idealization

The finite element discretization of the structure is well known. The fluid can be isolated from the structure, and the coupling can be achieved by introducing the structure-fluid interface pressures as the external forces acting on the structure. For excitation in the horizontal x-direction, the resulting set of equations in the general form is given by

$$[M] \{\ddot{w}\} + [C] \{\dot{w}\} + [K] \{w\} = R \qquad (5.5-1)$$

where

$$R = [M] \{1\} \ddot{x}(t) + \{L(Q)\} \qquad (5.5-2)$$

If the displacement of the structure is expressed in terms of interpolation functions as

$$\{w\} = \Sigma N_i w_i = [N] \{w\}, \qquad (5.5\text{-}3)$$

the pressure at the fluid nodes is specified by the interpolation function

$$\{p\} = \Sigma \bar{N}_i p_i = [\bar{N}] \{p\}. \qquad (5.5\text{-}4)$$

The pressure ('force') at the structure-fluid interface can be written as

$$\{P\} = [L] \{p\} \qquad (5.5\text{-}5)$$

where [L] is the coupling matrix, the elements of which can be evaluated using

$$\bar{L}_{ij} = \int_{\Gamma_s} N_i^T n \bar{N}_j \, d\Gamma \qquad (5.5\text{-}6)$$

in which n is the unit normal vector to the structure face, and the integration is carried over the interface region Γ_s. It should be noted that the matrix [L] has non-zero elements corresponding to interface nodes only.

5.5.2 Fluid Idealization

The linearized governing equation for the fluid pressure distribution including the compressibility effects and excluding viscous effects is given by Eq. 2.1-1.

The boundary condition due to the surface gravity waves is

$$(dp/dz) + (d^2p/dt^2)\frac{1}{g} = 0 \qquad (5.5\text{-}7)$$

where z is the coordinate in the vertical direction. At the bottom the boundary condition is

$$(dp/dz) = 0. \qquad (5.5\text{-}8)$$

The boundary condition at the interface, assuming the fluid to have the same velocity as the structure, is given as

$$\frac{\partial p}{\partial n} - \dot{s}_n = 0 \tag{5.5-9}$$

where $\dot{s}_n$ is the normal component of the structure velocity. At infinity, the radiation boundary condition is

$$(\partial p/\partial x) + \frac{1}{c}(\partial p/\partial t) = 0 \tag{5.5-10}$$

for the outgoing waves. In the above equation, p is the scattered component of the pressure.

Using the standard Galerkin's weighted residual approach, the governing equation and the boundary conditions will yield

$$[H]\{\ddot{p}\} + [T]\{\dot{p}\} + [G]\{p\} + \{P\} = 0. \tag{5.5-11}$$

The elements of the matrices in the above equation can be evaluated using

$$H_{ij} = \frac{1}{c^2} \int_\Omega N_i\, N_j\, d\Omega + \frac{1}{g} \int_{\Gamma_f} \bar{N}_i\, \bar{N}_n\, d\Gamma \tag{5.5-12}$$

where Ω is the fluid region, and Γ_f is the free surface,

$$T_{ij} = \frac{1}{c} \int_{\Gamma_\gamma} \bar{N}_i\, \bar{N}_j\, d\Gamma \tag{5.5-13}$$

where Γ_γ is the radiation boundary,

$$G_{ij} = \int_\Omega (\nabla \bar{N}_i)^t (\nabla \bar{N}_j)\, d\Omega \tag{5.5-14}$$

in which the subscripts 'i' and 'j' refer to the ith row and jth column of the matrix.

$$\{P\} = \int_{\Gamma_s} N_i\, \rho n^T\, \ddot{w}\, d\Gamma \tag{5.5-15}$$

$$= \rho[L]^T\{\ddot{w}\}. \tag{5.5-16}$$

5.5.3 Structure-Fluid Coupling

Combining equations (5.5-1) and (5.5-11) will yield the coupled structure-fluid equation (2.1-27). For any given external loading R, the above equation can be solved, either in time or the frequency domain, to yield the solution of the coupled system.

5.6 Mooring System

The overall design considerations for the mooring system have been reviewed by Nath [68]. Three different approaches for the determination of the dynamic behavior of the mooring system have been reviewed by Harlow, Harris, and Kehnemuyi [69]. Extrapolated hydrodynamic coefficients for the mooring system, based on the experimental values of similar structures, were tabulated. Theoretical analysis of the mooring system response to wind, wave, earthquake and tornado forces is described in Appendix 3K of Reference 70. The finite element formulation developed in the previous section can also be applied for the mooring system response analysis. If the mooring struts are designed as crushable tubes, the horizontal transmission of the earthquake input to the floating platform will be reduced significantly.

REFERENCES

1. Housner, G.W., "Dynamic Pressures on Accelerated Fluid Containers," Bulletin of the Seismological Society of America, Vol. 47, No. 1, Jan. 1957.

2. "Dynamic Pressure on Fluid Containers," Chapter 6 of Nuclear Reactors and Earthquakes, TID-7024, U.S. Atomic Energy Commission, Washington, D.C., Aug. 1963.

3. Abramson, H.N., editor, The Dynamic Behavior of Liquids in Moving Containers, NASA SP-106, National Aeronautics & Space Administration, Washington, D.C., 1966 (2nd Printing available from Southwest Research Institute, San Antonio, Texas, 1981).

4. Roberts, J.R., Eduardo, R.B., and Chen, P.Y., Slosh Design Handbook I, NASA CR-406, National Aeronautics and Space Administration, Washington, D.C., May 1966.

5. Kana, D.D., "Status and Research Needs for Prediction of Seismic Response in Liquid Containers," Nuclear Engineering and Design, 69, (1982), pp 205-221.

6. Dong, R.A., "Effective Mass and Damping of Submerged Structures," UCRL-52342, Lawrence Livermore Laboratory, April 1, 1978.

7. Healey, J.J., Wu, S.T., and Murga, M., "Structural Building Response Review," (Lawrence Livermore Laboratories), Report NUREG/CR-1423, Vol. 1, U.S. Nuclear Regulatory Commission, May 1980.

8. Clough, R.W., and Penzien, J., Dynamics of Structures, McGraw Hill Book Co., New York, 1975.

9. "Combining Modal Responses and Spatial Components in Seismic Response Analysis," U.S. NRC Reg. Guide 1.92, U.S. Nuclear Regulatory Commission, Washington, D.C., February 1976.

10. Bendat, J.S., and Piersol, A.G., Measurement and Analysis of Random Data, Wiley, New York, 1966.

11. Singh, M.P., "Seismic Design Input for Secondary Systems," Civil Engineering and Nuclear Power, Volume II, ASCE Preprint 3595 (1979).

12. Kaul, M.K., "Stochastic Characterizations of Earthquakes Through Their Response Spectrum," Earthquake Engineering and Structural Design, 6 (1978), pp 497-509.

13. Unruh, J.F., and Kana, D.D., "An Iterative Procedure for Generation of Consistent Power/Response Spectrum," Nuclear Engineering and Design 66, (1981), pp 427-435.

14. Niwa, A., "Seismic Behavior of Tall Liquid Storage Tanks," Earthquake Engineering Research Center Report UCB/EERC-78/04, February 1978.

15. Housner, G.W., and Haroun, M.A., "Vibration Tests of Full-Scale Liquid Storage Tanks," Proc. of 2nd U.S. National Conference on Earthquake Engineering, EERI, Stanford Univ., August 1979, pp 137-145.

16. Housner, G.W., and Haroun, M.A., "Seismic Design of Liquid Storage Tanks," J. of Tech. Council of ASCE, April, 1981.

17. Aslam, M., Godden, W.G., and Scalise, D.T., "Sloshing of Water in Torus Pressure-Suppression Pool of Boiling Water Reactors Under Earthquake Ground Motions," (Lawrence Berkeley Laboratories), Report NUREG/CR-1082, U.S. Nuclear Regulatory Commission, October 1979.

18. Kalnins, A., "Vibration of Fluid Filled Thin Shells," Paper B4/8, Proc. of 5th International Conference on Structural Mechanics in Reactor Technology, Berlin, August 1979.

19. Veletsos, A.S., and Yang, Y.Y., "Earthquake Response of Liquid Storage Tanks," Proc. of 2nd Annual ASCE Engineering Mechanics Specialty Conference, North Carolina State Univ., Raleigh, N.C., May 1977, pp 1-24.

20. Fischer, D., "Dynamic Fluid Effects in Liquid-Filled Flexible Cylindrical Tanks," Earthquake Engineering and Structural Dynamics, Vol. 7, 1979, pp 587-601.

21. Yang, J.Y., "Dynamic Behavior of Fluid-Tank Systems," Ph.D. Thesis, Rice University, Houston, Texas, March 1976.

22. Kana, D.D., and Craig, R., "Parametric Oscillations of a Longitudinally Excited Cylindrical Shell Containing Liquid," Journal of Spacecraft and Rockets, 5, 1, pp 13-21, January 1968.

23. Kana, D.D., and Dodge, F.T., "Design Support Modeling of Liquid Slosh in Storage Tanks Subject to Seismic Excitation," Proc. of ASCE Conference on Structural Design of Nuclear Plant Facilities, Vol. 1A, pp 307-337, December 1975.

24. Kana, D.D., "Seismic Response of Flexible Cylindrical Liquid Storage Tanks," Nuclear Engineering and Design, 52, 1979, pp 185-199.

25. Clough, D.P., and Clough, R.W., "Earthquake Simulator Studies of Cylindrical Tanks," Nuclear Engineering and Design, 46, (1978), pp 367-380.

26. Kana, D.D., "Liquid Slosh Response in a Horizontal Cylindrical Tank under Seismic Excitation," ASCE Conf. on Civil Engineering & Nuclear Power, Knoxville, Tennessee, Vol. VI Sept. 1980.

27. Balendra, T., et al, "Seismic Design of Flexible Cylindrical Liquid Storage Tanks," Earthquake Engineering and Structural Dynamics, Vol. 10, (1982), pp 477-496.

28. Aslam, M., Godden, W.G., and Scalise, D.T., "Earthquake Sloshing in Annular and Cylindrical Tanks," ASCE Journal of the Engineering Mechanics Division, Vol. 105, pp 371-389, June 1979.

29. Aslam, M., Godden, W.G., and Scalise, T., "Sloshing of Water in Annular Pressure-Suppression Pool of Boiling Water Reactors Under Earthquake Ground Motions," (Lawrence Berkeley Laboratories), Report NUREG/CR-1083, U.S. Nuclear Regulatory Commission, October 1979.

30. Fujita, K. and Shiraki, K., "Approximate Seismic Response Analysis of Self-Supported Thin Cylindrical Liquid Storage Tanks," Paper K 5/4, Proc. of 4th International Conference on Structural Mechanics in Reactor Technology, San Francisco, CA, August 1977.

31. Nakano, K., and Watabe, M., "Experimental Research on the Aseismic Characteristic of Spherical Steel Tank for Liquid Petroleum Gas," Proc. of 6th Joint Panel Conference of the U.S.-Japan, May 15-17, 1974, H.S. Lew, Ed., National Bureau of Standards, Washington, D.C.

32. Shames, J.H., "Mechanics of Fluids," McGraw-Hill Book Co., New York, N.Y. 1972.

33. Fritz, R.J., "The Effects of Liquids on the Dynamic Motions of Immersed Solids," Journal of Engineering for Industry, Trans. ASME, Feb. 1972.

34. Newmark, N.M., and Rosenblueth, E., Fundamentals of Earthquake Engineering, Prentice-Hall, Inc., 1971.

35. Clough, R.W., "Effects of Earthquakes on Underwater Structures," Proc. of 2nd World Conference on Earthquake Engineering, Tokyo, 1960.

36. Chandrasekaran, A.R., Saini, S.S., and Malhotra, M.M., "Virtual Mass of Submerged Structures," Journal of the Hydraulics Div., Proc. of the ASCE, May 1972.

37. Skop, R.A., Ramberg, S.E., and Ferer, K.M., "Added Mass and Damping Forces on Circular Cylinders," ASME Paper 76-Pet-3, 1976.

38. Stelson, T.E., and Mavis, F.T., "Virtual Mass and Acceleration in Fluids," ASCE Trans. Paper No. 2870, Vol. 122, 1957.

39. Amey, H.B., Jr., and Pomonik, G., "Added Mass and Damping of Submerged Bodies Oscillating Near the Surface," Offshore Technology Conference, 1976 Proc., Vol. 1, OTC 1557.

40. Chen, S.S., "Vibrations of a Group of Circular Cylindrical Structures in a Liquid," Trans. of the 3rd International Conference on Structural Mechanics in Reactor Technology, Vol. 1, Part D (Sept. 1975), pp 1-11.

41. Chen, S.S., Wambsgauss, M.W., and Jendrzejczyk, J.A., "Added Mass and Damping of a Vibrating Rod in Confined Viscous Fluids," Journal of Applied Mechanics, 98(2), June 1976.

42. Chen, S.S., "Vibration of Nuclear Fuel Bundles," Nuclear Engineering and Design, 35 (1975), pp 399-422.

43. Chen, S.S., "Dynamics of Heat Exchanger Tube Banks," ASME Paper No. 76-WA/FE-28 (1976), pp 1-7.

44. Chung, H., and Chen, S.S., "Vibration of a Group of Circular Cylinders in a Confined Fluid," ASME Paper No. 77-APM-16 (1977), pp 1-5.

45. Levy, S., Wilkinson, J.P.D., "Calculations of Added Water Mass Effect for Reactor System Components," Trans. of the 3rd International Conference on Structural Mechanics in Reactor Technology, September 1975.

46. Wu, R.W., Liu, L.K., Levy, S., "Dynamic Analysis of Multibody System Immersed in a Fluid Medium," Trans. 4th International Conference on Structural Mechanics in Reactor Technology, San Francisco, CA, August 1977.

47. Fritz, R.J., and Kiss, E., "The Vibration Response of a Cantilevered Cylinder Surrounded by an Annular Fluid," KAPL-M-6539, General Electric Company, February 1966.

48. Au-Yung, M.K., "Response of Reactor Internals to Fluctuating Pressure Forces," Nuclear Engineering and Design, 35 (1975), pp 361-375.

49. Lee, S.C., and Reddy, D.V., "Frequency Tuning of Offshoer Platforms by Liquid Sloshing," (To appear in Journal of Applied Ocean Research).

50. Veletsos, A.S., and Yang, J.Y., "Dynamics of Fixed-Base Liquid Storage Tanks," U.S.-Japan Seminar for Earthquake Engineering Research with Emphasis on Lifeline Systems, Tokyo, Japan, November 8-12, 1976.

51. Niwa, A., and Clough, R.W., "Buckling of Cylindrical Liquid-Storage Tanks Under Earthquake Loading," Earthquake Engineering and Structural Dynamics, Vol. 10 (1982), pp 107-122.

52. Yamamoto, T., and Nath, J.H., "Hydrodynamic Forces on Groups of Cylinders," Offshore Technology Conference Proc., 1976, Vol. 1, OTC 2449.

53. Shin, Y.S., Jendrzejczyk, J.A., and Wambsganss, M.W., "The Effect of Tube Support Interaction on the Dynamic Response of Heat Exchanger Tubes," Trans. 4th International Conference on Structural Mechanics in Reactor Technology, San Francisco, CA, August 1977.

54. Thangam Babu, P.V., and Reddy, D.V., "Existing Methodologies in the Design and Analysis of Offshore Nuclear Power Plants," (Extended Version), Nuclear Eng. and Design, 48, 167-205, 1978.

55. Thangam Babu, P.V., and Reddy, D.V., "Fluid-Structure Interaction Response Analysis of Floating Nuclear Plants Including the Effects of Mooring," Ocean Eng., Pergamon Press, Vol. 7, 707-741, 1980.

56. Thangam Babu, P.V., and Reddy, D.V., "A Numerical Integration Scheme for Solving Coupled Equations of Fluid-Structure Interaction Systems," Proc. Int. Conf. on Numerical Methods for Coupled Problems, Swansea Great Britain, Sept. 1981.

57. Thangam Babu, P.V. and Reddy, D.V., "Frequency Analysis of FNP Platforms using a High Precision Triangular Bending Element," Proc. Fourth Int. Conf. Structural Mechanics in Reactor Technology, San Francisco, Paper J 2/7, 12 pp, August 1977.

58. Arockiasamy, M., Reddy, D.V., Thangam Babu, P.V., and Haldar, A.K., "Stochastic Response of Floating Nuclear Plants to Seismic Forces," Proc. Fifth National Meeting of the Universities Council on Earthquake Eng. Research (UCEER), Cambridge, Mass., U.S.A., June 1978.

59. Reddy, D.V., Arockiasamy, M., Haldar, A.K., and Thangam Babu, P.V., "Response of an Offshore Nuclear Plant to Seismic Forces," Proc. Conf. B.N.E.S. Vibration in Nuclear Plant, Session 9, Keswick, U.K., 9 pp, May 1978.

60. Thangam Babu, P.F., Arockiasamy, M., and Reddy, D.V., "Coupled Fluid-Structure Interaction of Floating Platforms," Proc. Eighth Can. Congr. of Appl. Mech., Moncton, N.B., 633-634, 1981.

61. Thangam Babu, P.V., Reddy, D.V., and Arockiasamy, M., "Fluid-Structure Interaction of Floating Platforms," Proc. Sixth National Meeting of the Universities Council for Earthquake Engineering Research, Univ. Ill., Urbana, Ill., 215-217, May 1980.

62. Arockiasamy, M., Thangam Babu, P.V., and Reddy, D.V., "Probabilistic Seismic Fluid-Structure Interaction of Floating Nuclear Plants," Proc. Fifth Int. Conf. on Structural Mechanics in Reactor Technology, Berlin, August 1979.

63. Thangam Babu, P.V., "Dynamic Fluid-Structure Interaction Analysis of Floating Platforms," Ph.D. Thesis, Memorial University of Newfoundland, St. John's Nfld., Canada, July 1981.

64. Orr, R.S. and Dotson, C., "Offshore Nuclear Power Plants," Nuclear Engr. and Design, 25, 334-349, 1973.

65. Williamson, R.A., Kennedy, R.P., Bachman, R.E., and Chow, A.W., "Response of a Proposed Nuclear Powered Ship to Water-Transmitted Earthquake Vibrations," Technical Report, Nuclear and Systems Sciences Group, Holmes and Narver, Inc., 1975.

66. Hwang, Li-Sam and Tuck, E.O., "On the Oscillations of Harbours of Arbitrary Shape," Journal of Fluid Mechanics, Vol. 42, Part 3, pp 447-464, 1970.

67. Bettess, P., and Zienkiewicz, O.C., "Diffraction and Refraction on Surface Waves Using Finite and Infinite Elements," Int. J. for Num. Methods in Engineering, Vol. 11, pp 1271-1290, 1977.

68. Nath, J.H., "General Considerations for the Design of the Moorings for Offshore Nuclear Power Generating Stations," Dept. of Mech. Engineering and School of Oceanography, Oregon State Univ., Corvallis, Oregon NTIS-RLO-2227-T17-2, April 1974.

69. Harlow, E.H., Harris, F.R., and Kehnemuyi, M., "Mooring System for Atlantic Generating Station, Paper No. OTC2065, Vol. II, Proc. Offshore Tech. Conf., Houston, Texas, pp 335-348, 1974.

70. U.S. Atomic Energy Commission, Atlantic Generating Station 1 and 2 - Preliminary Safety Analysis Report, Docket Nos. STN-50-477 and STN-40-478, January 1977.

71. Thompson, W.T., Theory of Vibration with Applications, Prentice-Hall, Inc., New Jersey, 1981, p 276.

72. Roark, R.J., and Young, W.D., Formulas for Stress and Strain, McGraw-Hill, New York, 1975, p 185.

73. Haroun, M.A., "Vibration Studies and Tests of Liquid Storage Tanks," Earthquake Engineering and Structural Dynamics, Vol. II, pp 179-206, (1983).

74. Dean, R.G., Moran, G., Manley, R.N., Schmeltz, E.J., Omohundro, F., "Model Studies to Evaluate an Offshore Nuclear Power Plant Design," OTC Proceedings 1974, Paper No. 1986, pp 487-499.